THE HOLLOW EARTH

The Greatest Geographical Discovery in History

THE HOLLOW EARTH

The Greatest Geographical Discovery in History

Made by Admiral Richard E. Byrd in the
Mysterious Land Beyond the Poles —
The True Origin of the Flying Saucers

By Dr. Raymond Bernard, A.B., M.A., Ph.D.

Bell Publishing Company
New York

This edition is published by Bell Publishing Company,
a division of Crown Publishers, Inc.,
by arrangement with University Books, Inc.
    c d e f g h
BELL 1979 EDITION

Manufactured in the United States of America

Library of Congress Cataloging in Publication Data

Bernard, Raymond W
   The hollow Earth.

   Bibliography: pp. 16–17.
    1. Earth—Internal structure. 2. Byrd, Richard
Evelyn, 1888-1957. 3. Flying saucers. 4. Explorers—
United States—Biography. I. Title.
QE509.B47   1979   001.9   79-22645
ISBN 0-517-30793-6

DEDICATED

To the Future Explorers of the New World that exists beyond North and South Poles in the hollow interior of the Earth. Who will Repeat Admiral Byrd's historic Flight for 1,700 Miles beyond the North Pole and that of his Expedition for 2,300 Miles beyond the South Pole, entering a New Unknown Territory not shown on any map, covering an immense land area whose total size is larger than North America, consisting of forests, mountains, lakes, vegetation and animal life. The aviator who will be the first to reach this New Territory, unknown until Admiral Byrd first discovered it, will go down in history as a New Columbus and greater than Columbus, for while Columbus discovered a new continent, he will discover a New World.

PLANET SATURN

The planet Saturn is a world within a world, and maybe more. The inner world is flattened at the poles, and is 75,000 miles in diameter. If hollow, the earth could move round in it, and yet not be within 20,000 miles of its walls.

FOREWORD

Statements in this publication are recitals of scientific findings, known facts of physiology and references to ancient writings as they are found.

There are numerous authorities who have stated that flying saucers and other strange phenomena exist which scientists will not, or dare not admit. Few in the U.S. dare to state the whole truth and nothing but the truth about any subject (even if they know the truth). This is especially true of the "educated" *scientists* and men high and mighty. Governments would topple—money and credit would vanish. There would be utter chaos and high-ranking individuals would be ruined socially and economically!

Truth is such a rare quality, a stranger so seldom met in this civilization of fraud, that it is never received freely, but must fight its way into the world. There is not a public school which teaches the truth about Religion, Health, the Money-System, Politics of How to Buy and Sell etc.

We assume no responsibility for any opinions expressed (or implied) by the author. We have no authority to comment upon the opinions of the author. Books and folios are sold to be accepted or rejected, and the purpose of these writings is to dissipate darkness and to stir the minds of the people. Those who run may read—there are thousands of books which are far more enlightening than the "spoon-fed" news from the daily scandal sheets.

In November 1957, a world-famous physician and scientist died in a U.S. federal penitentiary where he had been imprisoned for resisting an unlawful injunction designed to stop his vital research, steal his discoveries, and kill the discoverer. That man was Wilhelm Reich, M.D. This was the culmination of more than 10 years of harassment and persecution at the hands of carefully concealed conspirators who used U.S. Federal Agencies and Courts to defraud the people of this earth and prevent them from knowing and utilizing crucial discoveries in physics, medicine, and sociology which could help bring about happiness and peace for all mankind which good men and women everywhere seek and work towards.

His "device" mentioned (in the injunction) was his "orgone energy accumulator," an invention of his which was hailed by the late Theodore P. Wolfe, M.D., as "the greatest single discovery in the history of medicine." The FDA completely evaded, avoided and banned (and burned) the published and documented verification of scores of reputable physicians and scientists THROUGHOUT THE WORLD, who duplicated, verified and published corroboration of the discoveries of Wilhelm Reich. These books claimed no cures of any kind. FDA said they constituted labeling (misbranding or mislabeling). His public experimental findings were revolutionary. They threatened at once, the existing commercial interests, especially the Drug Industry, the Power Companies, etc. The squeeze was on to *kill* this discovery, *kill it dead!*— just as Krebiozen and other cancer cures have been killed.

Therefore, I will not enter into any correspondence regarding this book—or the author. Whether you accept or reject the content of this book is your privilege. No one cares. If it awakens a responsive chord there are other books which may offer new knowledge for you (not taught in public schools or through the public mediums).

Robert Fieldcrest

UFOs AND GOVERNMENTAL SECRECY

The late Frank Edwards, courageously outspoken radio and television commentator, was quoted as saying: "Orders of secrecy (re UFOs) come from the *top*. The Air Force is simply the 'fall guy.'"

Edwards, an outstanding pioneer in the field of Ufology, carried on a vigorous crusade to expose official censorship and the withholding of facts relating to UFOs from the public. As a friend and acquaintance of Edwards for many years, I feel that he knew whereof he spoke; and I concur with the above-expressed opinion.

Recent revelations from various sources have served to substantially confirm his conclusion. It is with this thought in mind that the facts have been brought together for the reader's information and consideration.

For years many UFO investigators have felt that the Air Force was in possession of facts relating to UFOs which they were withholding from the public for reasons known only to themselves. Various opinions to this effect have appeared in print in recent years. A few of the most authoritative are quoted below.

In 1958, Bulkley Griffin, of the New Bedford, Mass. Standard-Times Washington bureau, wrote an excellent series of articles for that paper, one of which was entitled: "Pentagon's Censorship on UFOs is Effective." The following is a quotation from that article:

It has been asked what right has a military organization to monopolize control of the UFO situation and seriously interfere with the public's knowledge regarding it.

To this the Air Force has its own answer. Regulation 200-2 starts: "The Air Force investigation and analysis of UFOs over the United States are directly related to its responsibility for the defense of the United States." Later (in 200-2) the UFOs are listed as a "possible threat to the security of the United States."

But why the broad and persistent Air Force secrecy on UFOs? This essential question lacks a definite answer. The CIA, to an unknown extent, and the FBI, in individual cases, have taken an interest in the situation. Neither outfit is famed for unloading facts upon the public. The National Security Council operates in like fashion.

The staff of the Senate Permanent Subcommittee on Investigations has done some study on the UFO Air Force matter, and has decided not to investigate, and to hold no hearings open or closed. This is a victory for the Air Force, which is earnestly and discretely fighting any Congressional probe.

Late in 1958 another equally fine series of articles, by John Lester, appeared in the Newark Star Ledger. The following is a pertinent excerpt:

A news story revealing that government personnel are tracking mysterious objects was hailed yesterday as "one of the most important developments in the flying saucer problem."

Major Donald E. Keyhoe, Chairman of the National Investigations Committee on Aerial Phenomena, said:

"This newspaper's disclosures confirmed publicly what has been known to our investigations committee, i.e., that many highly qualified observers know that flying saucers are real and under intelligent control."

Keyhoe, a retired Marine officer, added that "this revelation by the Star-Beacon should help to bring down official censorship, which is keeping the truth from the public."

Behind the scenes, the Air Force and the Central Intelligence Agency are withholding the facts until they decide what to tell the American people.

One conclusion by a high Air Force intelligence agent is that the unidentified flying objects are interplanetary vehicles.

An item entitled "News Management" appeared in the January–February 1963 issue of Keyhoe's "UFO Investigator." In it, reference is made to a statement by Congressman John E. Moss, Chairman of the House Subcommittee on Government Information. This statement appeared in the Washington-World, in which Moss said:

Tighter controls are being used for greater manipulation of information on the Department of Defense level. . . . Information officers of the various services may be relegated to the status of a ventriloquist's dummy. The public does not have to prove its right to know.

Congressman Moss has told NICAP that his Committee is authorized to examine evidence showing the withholding of specific UFO reports or documents, though it is not empowered to investigate the general issue of UFO censorship.

Major Keyhoe, it may be added, has steadfastly adhered to the belief that flying saucers are real, prior to and following his first published article on the subject in the January 1950 issue of True Magazine.

Reference has been made to the CIA playing a role in the UFO secrecy picture. Further evidence of this fact is contained in a paragraph of the book "Inside Saucer Post 3–0 Blue," by Leonard H. Stringfield, formerly a well-known UFO investigator of Cincinnati, Ohio. On Page 42 of his book, Stringfield has this to say:

The statement itself came from Mr. A. D., of a certain top agency in Washington. Said A. D.: Yes, I did have a case for Federal Court. However, by use of the injunction, if necessary, he would prevent anyone from testifying in court concerning this book, because maximum security exists concerning the subject of the UFOs. My attorney therefore suggested we drop the case. . . .

Air Force saucer files, therefore, are . . . under the lock and key of "maximum security."

Several years after the publication of his book, Stringfield confirmed that the Mr. A. D. referred to was Allen Dulles, former director of the CIA. Stringfield's experience served to illustrate the serious interest which this agency has in the UFO matter.

At one time, along with many others, it was my opinion that it was the Air Force who were withholding the true facts about the UFOs. However, as a result of later developments, I no longer adhere to that conviction. One of the principal factors in my change of opinion was a conversation with the late Wilbert B. Smith, approximately two years before he passed away. During the course of this conversation, I asked Mr. Smith whether it was the Air Force or some other department of government that was keeping UFO information from the public. Mr. Smith replied that it was not the Air Force but "a small group very high up in the government." Upon further questioning, Mr. Smith refused to identify the group to which he had reference, and quickly led the conversation into other channels.

It was not until publication of the book "The Invisible Government" in May, 1964, that the apparent key to this mystery was at last provided. The book states:

The Special Group was created early in the Eisenhower years under the secret order 54/12. It was known in the innermost circle of the Eisenhower Administration as the "54/12 Group" and is still so called by a few insiders. . . . It has operated for a decade as the hidden power center of the Invisible Government. . . . The Special Group has operated in an atmosphere of secrecy exceeding that of any other branch of the United States Government. . . . CIA men generally have the Special Group in mind when they insist that the agency has never set policy, but has only acted on higher authority.

Newsweek of June 22nd, 1964, carried a review of "The Invisible Government" which stated: "One of their major revelations is the existence of the Special Group '54/12,' a hitherto Classified adjunct of the National Security Council, specifically charged by the President with ruling on special operations. Practically speaking, there are no higher-level figures imaginable than the composition of '54/12.' " Names of members of the Group follow at this point.

Was the "54/12 Group" the "small group very high up in Government" to which Wilbert Smith had reference? In view of the preceding facts, it would seem that the logical answer to this question could very well be in the affirmative! And, if this assumption is correct, it could provide the answer as to why all attempts to date to obtain open hearings on the UFO matter have met with failure.

One can only hope that the public will eventually realize their inalienable right to know the truth about the UFOs. Or can a top-flight "Special Group" thwart their efforts? Only time will tell!

References:

"The Invisible Government" by David Wise and Thomas B. Ross—Random House.

"Who Rules America?" by John McConaughy—Longmans, Green & Co.

"The Iron Curtain Over America" by John Beaty—Wilkinson Pub. Co.

"Freedom or Secrecy" by James R. Wiggins—Oxford University Press.

"The Right to Know" by Kent Cooper—Farrar, Straus & Cudahy.

"Managed News" by Hanson W. Baldwin—Atlantic Magazine, April, 1964.

"News Management in Washington" by Ben H. Bag-
dikian—Sat. Eve. Post, 4/20/63.
"CIA—The Battle for Secret Power" by Stewart Alsop—
Sat. Eve. Post, 8/3/63.

LANDMARKS IN THE HISTORY OF REAR ADMIRAL RICHARD E. BYRD'S GREAT GEOGRAPHICAL DISCOVERY OF NEW UNKNOWN LAND AREAS WITHIN THE POLAR CONCAVITIES BEYOND NORTH AND SOUTH POLES

DECEMBER, 1929: "The memorable December 12th discovery of heretofore unknown land beyond the South Pole by Capt. Sir George Hubert Wilkins, demands that science change the concept it has had for the past four hundred years concerning the southern contour of the earth."

—Dumbrova, Russian Explorer

FEBRUARY, 1947: "I'd like to see that *land beyond the* (North) *Pole*. That area beyond the Pole is the center of the *Great Unknown*."

—Rear Admiral Richard E. Byrd (United States Navy), before his seven-hour flight of 1,700 miles beyond the North Pole. (Author's note: Admiral Byrd did not fly 1,700 miles *across* the North Pole to the other side of the earth, a frozen icy waste, like the region from which he came—but flew *beyond* the Pole into the polar opening leading to the hollow interior of the earth, traversing an iceless region of mountains, lakes, rivers, green vegetation and animal life.)

NOVEMBER, 1955: *"This is the most important expedition in the history of the world."*

       —Rear Admiral Richard E. Byrd, before departing to explore land beyond the South Pole.

JANUARY, 1956: "On January 13th, members of the United States expedition accomplished a flight of 2,700 miles from the base at McMurdo Sound, which is 400 miles west of the South Pole, and penetrated a land extent of 2,300 *miles* beyond the Pole."

       —Radio announcement from Byrd's Antarctic expedition, confirmed by the American press of February 5, 1956.

MARCH, 1956: "The present expedition has opened up *a vast new territory*."

       —Rear Admiral Byrd, on March 13, 1956, when he returned from his South Polar expedition.

. . . "that enchanted continent in the sky, *land of everlasting mystery!*"

       —Rear Admiral Byrd, before his death. (Author's note: Byrd here enigmatically refers to the new unknown land area he discovered beyond North and South Poles, within the polar openings, which, due to the well known polar phenomenon of "sky mirror," whereby land areas below are mirrored in the sky, refers here to the new land areas he discovered beyond both North and South Poles and beheld as "an enchanted continent in the sky.")

WHAT THIS BOOK SEEKS TO PROVE

1. That the Earth is hollow and is not a solid sphere as commonly supposed, and that its hollow interior communicates with the surface by two polar openings.

2. That the observations and discoveries of Rear Admiral Richard E. Byrd of the United States Navy, who was the first to enter into the polar openings, which he did for a total distance of 4,000 miles in the Arctic and Antarctic, confirm the correctness of our revolutionary theory of the Earth's structure, as do the observations of other Arctic explorers.

3. That, according to our geographical theory of the Earth being concave, rather than convex, at the Poles, where it opens into its hollow interior, the North and South Poles have never been reached because they do not exist.

4. That the exploration of the unknown New World that exists in the interior of the Earth is much more important than the exploration of outer space; and the aerial expeditions of Admiral Byrd show how such exploration may be conducted.

5. That the nation whose explorers first reach this New World in the hollow interior of the Earth, which has a land area greater than that of the Earth's surface, which may be done by retracing Admiral Byrd's flights beyond the hypothetical North and South Poles, into the

Arctic and Antarctic polar openings, will become the greatest nation in the world.

6. That there is no reason why the hollow interior of the earth, which has a warmer climate than on the surface, should not be the home of plant, animal and human life; and if so, it is very possible that the mysterious flying saucers come from an advanced civilization in the hollow interior of the earth.

7. That, in event of a nuclear world war, the hollow interior of the earth will permit the continuance of human life after radioactive fallout exterminates all life on the Earth's surface; and will provide an ideal refuge for the evacuation of survivors of the catastrophe, so that the human race may not be completely destroyed, but may continue.

FOREWORD

It is the purpose of this book to present scientific evidence to prove that the Earth, rather than being a solid sphere with a fiery center of molten metal, as generally supposed, is really hollow, with openings at its poles. Also, in its hollow interior exists an advanced civilization which is the creator of the flying saucers.

The theory of a hollow earth was first worked out by an American writer, William Reed, in 1906, and later extended by another American, Marshall B. Gardner, in 1920. In 1959, F. Amadeo Giannini wrote the first book on the subject since Gardner's, and in the same year, Ray Palmer, editor of "Flying Saucers" magazine, applied the theory to provide a logical explanation for the origin of the flying saucers.

The theories of Reed and Gardner found confirmation in the Arctic and Antarctic expeditions of Rear Admiral Richard E. Byrd in 1947 and 1956 respectively, which penetrated for 1,700 miles beyond the North Pole and 2,300 miles beyond the South Pole, into new unknown, iceless territory not recorded on a map, extending inside the polar depressions and openings that lead to the hollow interior of the Earth. The true significance of Admiral Byrd's great discoveries was hushed up soon after he sent his radio report from his plane, and was not given the attention it deserves until Giannini and Palmer publicized the matter. We shall explain below the reason why this information was not permitted to reach the public, though it concerns one of the greatest geographical discoveries in history, far greater than Columbus's discovery of America, for while Columbus discovered a new continent, Byrd discovered a New World with a land area equal to or greater than the entire land surface of the Earth.

Admiral Byrd's discovery is today a leading international top secret, and it has been so since it was first made in 1947. After Byrd made his radio announcement from his plane and after a brief press notice, all subsequent news on the subject was carefully suppressed by government agencies. There was an important reason for this. Before he left on his seven hour flight from his Arctic base over iceless land *beyond* the North Pole (leading to the interior of the Earth), Admiral Byrd said: "I would like to see that land *beyond* the Pole. That area beyond the Pole is the center of the Great Unknown."

Admiral Byrd did not cross over the North Pole and travel 1,700 miles south on its other side. If he did, he would enter icebound territory. Instead he entered a land with a warmer climate, free from ice and snow, consisting of forests, mountains, lakes, green vegetation and animal life. This new *unknown* land over which he flew for 1,700 miles, which was not on any map, existed *inside* the polar opening leading to the hollow interior of the Earth, where it is warmer than on its outside, which is here a land of ice and snow.

Why then did not the United States send new air expeditions to the land discovered by Admiral Byrd, in an effort to fully explore it and determine its extent? Why was such an important discovery completely forgotten? It is like Columbus discovering America and then nothing further being done about it and no subsequent trips of exploration to the Western continent made by Europeans. Why the apathy?

The explanation is evident. If Admiral Byrd made such a momentous discovery, undoubtedly the greatest one in history, of a new unknown land area of undetermined extent, over which his expedition flew for a total of 4,000 miles at each pole, and which area is probably as wide as it is long and, since Byrd turned back before reaching its end, is probably much larger than an area 4,000 miles square, then it would be in the interest of the U.S. Government to keep this discovery secret, so that other nations do not learn about it and claim this territory for themselves.

It seems that news of Admiral Byrd's great discovery did not reach the Soviet Union, or else the Soviet Union knows about this new land area not contained on any map, but has adopted the same policy of silence and secrecy.

If the Soviet Union knew about the discovery it would surely send fleets of atomic-powered submarines, ice-breakers and airplanes into this unknown territory beyond the Pole and be the first to explore it and claim it for its own. To prevent this was probably the reason that the news about Admiral Byrd's great discovery was hushed and suppressed ever since it was first released. However, since the secret has already been released and broadcast by Giannini, Palmer and others, and is public knowledge, it can no longer properly be called a secret.

It is hoped that a serious expedition will be made by a neutralist, peace-loving nation like Brazil into this New World beyond the Poles and establish contact with the advanced civilization that exists there, whose flying saucers are evidence of their superiority over us in scientific development. Perhaps this elder wiser race may save us from our doom, preventing a future nuclear war and enabling us to establish a New Age on earth, an age of permanent peace, with all nuclear weapons outlawed and destroyed by a world government representing all the peoples on earth.

The ring or hollow shell nebula in Lyra was evolved from masses of nebulous matter, showing the polar opening and central sun, which will finally evolve itself into a new planet. (*Photographed at Lick Observatory*)

A spiral nebula showing the central nucleus projecting masses of nebulous matter which forms a ring or wall around this central body, as clearly shown in the accompanying reproduction of a ring nebula. (*Photographed at Yerkes Observatory, January 3, 1912*)

Chapter I

ADMIRAL BYRD'S
EPOCH-MAKING DISCOVERY

The Greatest Geographical Discovery in Human History

> "That enchanted Continent in the Sky, Land of
> Everlasting Mystery!"
> "I'd like to see that land beyond the (North) Pole.
> *That area beyond the Pole is the Center of the Great
> Unknown!*"
> —Rear Admiral Richard E. Byrd

The above two statements by the greatest explorer in
modern times, Rear Admiral Richard E. Byrd of the
United States Navy, cannot be understood or make any
sense according to old geographical theories that the
earth is a solid sphere with a fiery core, on which both
North and South Poles are fixed points. If such was the
case, and if Admiral Byrd flew for 1,700 and 2,300 miles
respectively *across* North and South Poles, to the icy and
snowbound lands that lie on the other side, whose geography is fairly well known, it would be incomprehensible
for him to make such a statement, referring to this territory on the other side of the Poles as "the great unknown." Also, he would have no reason to use such a
term as "Land of Everlasting Mystery." Byrd was not a
poet, and what he described was what he observed from
his airplane. During his Arctic flight of 1,700 miles *beyond* the North Pole he reported by radio that he saw

below him, not ice and snow, but land areas consisting of mountains, forests, green vegetation, lakes and rivers, and in the underbrush saw a strange animal resembling the mammoth found frozen in Arctic ice. Evidently he had entered a warmer region than the icebound Territory that extends from the Pole to Siberia. If Byrd had this region in mind he would have no reason to call it the "Great Unknown," since it could be reached by flying across the Pole to the other side of the Arctic region.

The only way that we can understand Byrd's enigmatical statements is if we discard the traditional conception of the formation of the earth and entertain an entirely new one, according to which its Arctic and Antarctic extremities are not convex but concave, and that Byrd entered into the polar concavities when he went *beyond* the Poles. In other words, he did not travel *across* the Poles to the other side, but entered into the polar concavity or depression, which, as we shall see later in this book, opens to the hollow interior of the earth, the home of plant, animal and human life, enjoying a tropical climate. This is the "Great Unknown" to which Byrd had reference when he made this statement—and *not* the ice- and snow-bound area on the other side of the North Pole, extending to the upper reaches of Siberia.

The new geographical theory presented in this book, for the first time, makes Byrd's strange, enigmatical statements comprehensible and shows that the great explorer was not a dreamer, as may appear to one who holds on to old geographical theories. Byrd had entered an entirely new territory, which was "unknown" because it was not on any map, and it was not on any map because all maps have been made on basis of the belief that the earth is spherical and solid. Since nearly all lands on this solid sphere have been explored and recorded by polar explorers, there could not be room on such maps for the territory that Admiral Byrd discovered, and which he called the "Great Unknown"—unknown because not on any map. It was an area of land as large as North America!

This mystery can only be solved if we accept the basic conception of the earth's formation presented in this book and supported by the observations of Arctic explorers which will be cited here. According to this new revolutionary conception, the earth is not a solid sphere, but is *hollow*, with openings at the Poles, and *Admiral Byrd entered these openings* for a distance of some 4,000 miles during his 1947 and 1956 Arctic and Antarctic expeditions. The "Great Unknown" to which Byrd referred was the iceless land area inside the polar concavities, opening to the hollow interior of the earth. If this conception is correct, as we shall attempt to prove, then both North and South Poles cannot exist, since they would be in midair, in the center of the polar openings, and would not be on the earth's surface. This view was first presented by an American writer, William Reed, in a book, "Phantom of the Poles," published in 1906 soon after Admiral Peary claimed to have discovered the North Pole and denying that he really did. In 1920 another book was published, written by Marshall Gardner, called "A Journey to the Earth's Interior or Have the Poles Really Been Discovered?," making the same claim. Strangely, Gardner had no knowledge of Reed's book and came to his conclusions independently. Both Reed and Gardner claimed that the earth is hollow, with openings at the poles and that in its interior lives a vast population of millions of inhabitants, composing an advanced civilization. This is probably the "Great Unknown" to which Admiral Byrd referred.

To repeat, Byrd could not have had any part of the Earth's known surface in mind when he spoke of the "Great Unknown," but rather a new, hitherto unknown land area, free from ice and snow, with green vegetation, forests and animal life, that exists nowhere on the Earth's surface but inside the polar depression, receiving its heat from its hollow interior, which has a higher temperature than the surface, with which it communicates. Only on the basis of this conception can we understand Admiral Byrd's statements.

In January, 1956, Admiral Byrd led another expedition to the Antarctic and there penetrated for 2,300 miles *beyond* the South Pole. The radio announcement at this time (January 13, 1956) said: "On January 13, members of the United States expedition penetrated a land extent of 2,300 miles *beyond* the Pole. The flight was made by Rear Admiral George Dufek of the United States Navy Air Unit."

The word "beyond" is very significant and will be puzzling to those who believe in the old conception of a solid earth. It would then mean the region on the other side of the Antarctic continent and the ocean beyond, and would not be "a vast new territory" (not on any map), nor would his expedition that found this territory be "the most important expedition in the history of the world." The geography of Antarctica is fairly well known, and Admiral Byrd has not added anything significant to our knowledge of the Antarctic continent. If this is the case, then why should he make such apparently wild and unsupported statements—especially in view of his high standing as a rear admiral of the U.S. Navy and his reputation as a great explorer?

This enigma is solved when we understand the new geographical theory of a Hollow Earth, which is the only way we can see sense in Admiral Byrd's statements and not consider him as a visionary who saw mirages in the polar regions or at least imagined he did.

After returning from his Antarctic expedition on March 13, 1956, Byrd remarked: "The present expedition has opened up a vast new land." The word "land" is very significant. He could not have referred to any part of the Antarctic continent, since none of it consists of "land" and all of it of ice, and, besides, its geography is fairly well known and Byrd did not make any noteworthy contribution to Antarctic geography, as other explorers did, who left their names as memorials in the geography of this area. If Byrd discovered a vast new area in the Antarctic, he would claim it for the United States Government and it would be named after him, just as would be the case if his 1,700 mile flight beyond the North Pole was over the earth's surface between the Pole and Siberia.

But we find no such achievements to the credit of the great explorer, nor did he leave his name in Arctic and Antarctic geography to the extent that his statements about discovering a new vast land area would indicate. If his Antarctic expedition opened up a new immense region of ice on the frozen continent of Antarctica, it would not be appropriate to use the word "land," which means an iceless region similar to that over which Byrd flew for 1,700 miles *beyond* the North Pole, which had green vegetation, forests and animal life. We may therefore conclude that his 1956 expedition for 2,300 miles beyond the South Pole was over similar iceless territory not recorded on any map, and not over any part of the Antarctic continent.

The next year, in 1957, before his death, Byrd called this *land beyond* the South Pole (not "ice" on the other side of the South Pole) "that enchanted continent in the sky, land of everlasting mystery." He could not have used this statement if he referred to the part of the icy continent of Antarctica that lies on the other side of the South Pole. The words "everlasting mystery" obviously refer to something else. They refer to the warmer territory not shown on any map that lies inside the South Polar Opening leading to the hollow interior of the Earth.

The expression "that enchanted continent in the sky" obviously refers to a land area, and not ice, mirrored in the sky which acts as a mirror, a strange phenomenon observed by many polar explorers, who speak of "the island in the sky" or "water sky," depending on whether the sky of polar regions reflects land or water. If Byrd saw the reflection of water or ice he would not use the word "continent," or call it an "enchanted" continent. It was "enchanted" because, according to accepted geographical conceptions, this continent which Byrd saw reflected in the sky (where water globules act as a mirror for the surface below) could not exist.

We shall now quote from Ray Palmer, editor of "Flying Saucers" magazine and a leading American expert on flying saucers, who is of the opinion that Admiral Byrd's discoveries in the Arctic and Antarctic regions offer an explanation of the origin of the flying saucers, which, he believes, do not come from other planets, but from the hollow interior of the earth, where exists an advanced civilization far in advance of us in aeronautics, using flying saucers for aerial travel, coming to the outside of the earth through the polar openings. Palmer explains his views as follows:

"How well known is the Earth? Is there any area on Earth that can be regarded as a possible origin of the flying saucers? There are two. The two major areas of importance are the Antarctic and the Arctic.

"Admiral Byrd's two flights over both Poles prove that there is a 'strangeness' about the shape of the Earth in both polar areas. Byrd flew to the North Pole, but did not stop there and turn back, but went for 1,700 miles beyond it, and then retraced his course to his Arctic base (due to his gasoline supply running low). As progress was made beyond the Pole point, iceless land and lakes, mountains covered with trees, and even a monstrous animal, resembling the mammoth of antiquity, was seen moving through the underbrush; and all this was reported via radio by the plane occupants. For almost all of the 1,700 miles, the plane flew over land, mountains, trees, lakes and rivers.

"What was this unknown land? Did Byrd, in traveling due north, enter into the hollow interior of the Earth through the north polar opening? Later Byrd's expedition went to the South Pole and after passing it, went 2,300 miles beyond it.

"Once again we have penetrated an unknown and mysterious land which does not appear on today's maps. And once again we find no announcement beyond the initial announcement of the achievement [due to official suppression of news about it—Author]. And, strangest of all, we find the world's millions absorbing the announcements and registering a complete blank in so far as curiosity is concerned.

"Here, then, are the facts. *At both poles exist unknown and vast land areas, not in the least uninhabitable, extending distances which can only be called tremendous because they encompass an area bigger than any known continental area!* The North Pole Mystery Land seen by Byrd and his crew is at least 1,700 miles across its traversed direction, and cannot be conceived to be merely a narrow strip. It is an area perhaps as large *as the entire United States!*

"In the case of the South Pole, the land traversed beyond the Pole included an area as big as North America plus the south polar continent.

"The flying saucers could come from these two unknown lands 'beyond the Poles.' It is the opinion of the editors of 'Flying Saucers' magazine that the existence of these lands cannot be disproved by anyone, consider-

ing the facts of the two expeditions which we have outlined."

If Rear Admiral Byrd claimed that his south polar expedition was "the most important expedition in the history of the world," and if, after he returned from the expedition, he remarked, "The present expedition has
opened up a new vast land," it would be strange and inexplicable how such a great discovery of a new land area
as large as North America, comparable to Columbus's
discovery of America, should have received no attention
and have been almost totally forgotten, so that nobody
knew about it, from the most ignorant to the most
learned.

The only rational answer to this mystery is that, after
the brief announcement in the American press based on
Admiral Byrd's radio report, further publicity was suppressed by the Government, in whose employ Byrd was
working, and which had important political reasons why
Admiral Byrd's historic discovery should not be made
known to the world. For he had discovered two unknown
land areas measuring a total of 4,000 miles across and
probably *as large as both the North and South American
continents*, since Byrd's planes turned back without
reaching the end of this territory not recorded on any
map. Evidently, the United States Government feared
that some other government may learn about Byrd's discovery and conduct similar flights, going much further
into it than Byrd did, and perhaps claiming this land area
as its own.

Commenting on Byrd's statement, made in 1957
shortly before his death, in which he called the new ter-

ritory he discovered beyond the Poles "that enchanted
continent in the sky" and "land of everlasting mystery,"
Palmer says:

"Considering all this, is there any wonder that all the
nations of the world suddenly found the south polar region (particularly) and the north polar region so in-
tensely interesting and important, and have launched explorations on a scale actually tremendous in scope?"

Palmer concludes that this new land area that Byrd
discovered and which is not on any map, exists *inside* and
not outside the earth, since the geography of the outside
is quite well known, whereas that of the inside (within
the polar depression) is "unknown." For that reason
Byrd called it the "Great Unknown."

After discussing the significance of the use of the term
"beyond" the Pole by Byrd instead of "across" the Pole
to the other side of Arctic or Antarctic regions, Palmer
concludes that what Byrd referred to was an *unknown*
land area *inside* the polar concavity and connecting with
the warmer interior of the Earth, which accounts for its
green vegetation and animal life. It is "unknown" because it is not on the Earth's outer surface and hence is
not recorded on any map. Palmer writes:

"In February of 1947, Admiral Richard E. Byrd, the
one man who has done the most to make the North Pole
a known area, made the following statement: 'I'd like to
see the land *beyond* the Pole. That area beyond the Pole
is the *center of the Great Unknown.*'

"Millions of people read this statement in their daily
newspapers. And millions thrilled at the Admiral's subsequent flight to the Pole and to a point 1,700 miles be-

yond it. Millions heard the radio broadcast description of the flight, which was also published in newspapers.

"What land was it? Look at your map. Calculate the distance from all the known lands we have previously mentioned (Siberia, Spitzbergen, Alaska, Canada, Finland, Norway, Greenland and Iceland). A good portion of them are well within the 1,700 mile range. But none of them are within 200 miles of the Pole. Byrd flew over no known land. He himself called it 'the great unknown.' And great it is indeed! For after 1,700 miles over land, he was forced by gasoline supply shortage to return, and he had not yet reached the end of it! He should have been back to 'civilization.' But he was not. He should have seen nothing but ice-covered ocean, or at the very most, partially open ocean. Instead he was over *mountains covered with forest!*

"Forests!

"Incredible! The northernmost limit of the timberline is located well down into Alaska, Canada and Siberia. North of that line, no tree grows! All around the North Pole, the tree does not grow within 1,700 miles of the Pole!

"What have we here? We have the well-authenticated flight of Admiral Richard E. Byrd to a land *beyond* the Pole that he so much wanted to see, because it was the *center of the great unknown,* the center of mystery. Apparently, he had his wish gratified to the fullest, yet today, nowhere is this mysterious land mentioned. Why? Was that 1947 flight fiction? Did all the newspapers lie? Did the radio from Byrd's plane lie?

"No, Admiral Byrd did fly beyond the Pole.

"Beyond?

"What did the Admiral mean when he used that word? How is it possible to go 'beyond' the Pole? Let us consider for a moment. Let us imagine that we are transported by some miraculous means to the exact point of the North Magnetic Pole. We arrive there instantaneously, not knowing from which direction we came. And all we know is that we are to proceed from the Pole to Spitzbergen. But where is Spitzbergen? Which way do we go? South of course! But which South? All directions from the North Pole are south!

"This is actually a simple navigational problem. All expeditions to the Pole, whether flown, or by submarine, or on foot, have been faced with this problem. Either they must retrace their steps, or discover which southerly direction is the correct one to their destination, wherever it has been determined to be. The problem is solved by making a turn in any direction, and proceeding approximately 20 miles. Then we stop, measure the stars, correlate with our compass reading (which no longer points straight down, but toward the North Magnetic Pole), and plot our course on the map. Then it is a simple matter to proceed to Spitzbergen by going south.

"Admiral Byrd did not follow this traditional navigational procedure. When he reached the Pole, he continued for 1,700 miles. To all intents and purposes, he continued on a northerly course, after crossing the Pole. And weirdly, it stands on record that he succeeded, or he would not see that 'land beyond the Pole,' which to this day, if we are to scan the records of newspapers, books,

radio, television and word of mouth, has never been re-
visited!

"That land, on today's maps, *cannot exist*. But since it
does, we can only conclude that today's maps are incor-
rect, incomplete and do not represent a true picture of
the Northern Hemisphere.

"Having thus located a great land mass in the North,
not on any map today, a land which is the *center of the
great unknown,* which can only be construed to imply
that the 1,700 mile extent traversed by Byrd is only a portion of it."

Such an important discovery, which Byrd called "the
most important" in the history of the world, should have
been known to everyone, if information about it was not
suppressed to such an extent that it was almost completely forgotten until Giannini mentioned it in his book
"Worlds Beyond the Poles," published in New York in
1959. Similarly, Giannini's book, for some strange reason,
was not advertised by the publisher and remained unknown.

At the end of the same year, 1959, Ray Palmer, editor
of "Flying Saucers" magazine, gave publicity to Admiral
Byrd's discovery, about which he learned in a copy of
Giannini's book he read. He was so much impressed that
in December of that year he published this information
in his magazine, which was for sale on newsstands
throughout the United States. Then followed a series of
strange incidents, indicating that secret forces were at
work to prevent the information contained in the December issue of "Flying Saucers" magazine, derived from

Giannini's book, from reaching the public. Who are these secret forces that have a special reason to suppress the release of information about Admiral Byrd's great discovery of new land areas not on any map? Obviously, they are the same forces that suppressed news release of information, except for a brief press notice, after Byrd made his great discovery and before Giannini published the first public statement about it in many years, in 1959, twelve years after the discovery was made.

Palmer's announcement of Byrd's discoveries in the Arctic and Antarctic was the first large scale publicity since the time they were made and briefly announced, and so much more significant than Giannini's quotations and statements in his book that was not properly advertised and enjoyed a limited sale. For this reason, soon after the December, 1959 issue of "Flying Saucers" was ready to mail to subscribers and placed on newsstands, it was mysteriously removed from circulation—evidently by the same secret forces that suppressed the public release of this information since 1947. When the truck arrived to deliver the magazines from the printer to the publisher, no magazines were found in the truck! A phone call by the publisher (Mr. Palmer) to the printer resulted in his not finding any shipping receipt proving shipment to have been made. The magazines having been paid for, the publisher asked that the printer return the plates to the press and run off the copies due. But, strangely, the plates were not available, and were so badly damaged that no re-printing could be made. But where were the thousands of magazines that had been printed

and had mysteriously vanished? Why was there no shipping receipt? If it was lost and the magazines were sent
to the wrong address, they would turn up somewhere.
But they did not.

As a result, 5000 subscribers did not get the magazine.
One distributor who received 750 copies to sell on his
newsstand was reported missing, and 750 magazines disappeared with him. These magazines were sent to him
with the request that they be returned if not delivered.
They did not come back. Since the magazine disappeared
completely, several months later it was republished and
sent to subscribers.

What did this magazine contain that caused it to be
suppressed in this manner—by invisible and secret
forces? It contained a report of Admiral Byrd's flight beyond the North Pole in 1947, knowledge concerning
which was previously suppressed except for mention of
it in Giannini's book, "Worlds Beyond the Poles." The
December, 1959 issue of "Flying Saucers" was obviously
considered as dangerous by the secret forces that had a
special reason to withhold this information from the
world and keep it secret. In this issue of "Flying
Saucers," the following statements were quoted from
Giannini's book:

"Since December 12, 1929, U.S. Navy polar expeditions have determined the existence of indeterminable
land extent beyond the Pole points.

"On January 13, 1956, as this book was being prepared, a U.S. air unit penetrated to the extent of 2,300
miles beyond the assumed South Pole end of the earth.

That flight was always over land and water and ice. For very substantial reasons, the memorable flight received negligible press notice.

"The United States and more than thirty other nations prepared unprecedented polar expeditions for 1957–1958 to penetrate land now proved to extend beyond both Pole points. My original disclosure of then unknown land beyond the Poles, in 1926–1928, was captioned by the press as 'more daring than anything Jules Verne ever conceived.' Then Giannini quoted the following statements by Admiral Byrd we have presented above:

"1947: February. 'I'd like to see that land *beyond the Pole*. That area beyond the Pole is the center of the great unknown.'—Rear Admiral Richard E. Byrd, United States Navy, before his seven-hour flight over land beyond the North Pole.

"1956: January 13. 'On January 13, members of the United States expedition accomplished a flight of 2,700 miles from the base at McMurdo Sound, which is 400 miles west of the South Pole, and penetrated a land extent of 2,300 miles beyond the Pole.'—Radio announcement, confirmed by press of February 5.

"1956: March 13. 'The present expedition has opened up a vast new land'—Admiral Byrd, after returning from the Land beyond the South Pole.

"1957: 'That enchanting continent in the sky, land of everlasting mystery'—Admiral Byrd."

No attention was given by the scientific world to Giannini's book. The strange and revolutionary geographical theory it presented was ignored as eccentric rather than scientific. Yet Admiral Byrd's statements only make sense

if some such conception of the existence of "land beyond
the Poles," as Giannini claimed to exist, is accepted.
Giannini writes:

"There is no physical end of the Earth's northern and
southern extremities. The Earth cannot be circumnavigated north and south within the meaning of the word,
'circumnavigate.' However, certain 'round the world'
flights have contributed to the popular misconception
that the Earth has been circumnavigated north and
south.

" 'Over the North Pole,' with return to the North
Temperate Zone areas, without turning around, can
never be accomplished because there is no northern end
of the Earth. The same conditions hold true for the
South Pole.

"The existence of worlds beyond the Poles has been
confirmed by United States Naval exploration during the
past thirty years. The confirmation is substantial. The
world's eldest explorer, Rear Admiral Richard Evelyn
Byrd commanded the government's memorable expedition into that endless land beyond the South Pole. Prior
to his departure from San Francisco he delivered the momentous radio announcement, 'This is the most important expedition in the history of the world.' The subsequent January 13, 1956 penetration of land beyond the
Pole to the extent of 2,300 miles proved that the Admiral
had not been exaggerating."

Commenting on Giannini's statements about the im-
possibility of going straight north over the North Pole
and reaching the other side of the world, which would be
the case if the Earth was convex, rather than concave, at

the Pole, Palmer writes in his magazine, "Flying
Saucers:"

"Many of the readers stated that commercial flights
continually cross the Pole and fly to the opposite side of
the Earth. This is not true, and though the Airline of-
ficials themselves, when asked, might say that they do, it
is not literally true. They do make navigational maneuvers which automatically eliminate a flight beyond
the Pole in a straight line, in every sense. Ask the pilots
of these polar flights. And when they come to the exact
point, name one trans-polar flight on which you can buy
a ticket which actually *crosses* the North Pole.

"Examining the route of flights across the North Polar
area we always find that they go around the Pole or to
the side of it and never directly across it. This is strange.
Surely a flight advertised as passing directly over the
North Pole would attract many passengers who would
like to have that experience. Yet, strangely, no airline offers such a flight. Their air routes always pass on one side
of the Pole. Why? Is it not possible that if they went
straight across the Pole, instead of landing on the opposite side of the Earth, the plane would go to that *land
beyond the Pole*, 'the center of the Great Unknown,' as
Admiral Byrd called it?"

Palmer suggests that such an expedition that travels
directly north and continues north after reaching the
North Pole point (which he believes is in the center of
the polar concavity and not on solid land at all) should
be organized, retracing Admiral Byrd's route and continuing onward in the same direction, until the hollow
interior of the earth is reached. This, apparently, was

never done, in spite of the fact that the United States Navy, in its archives, has a record of Admiral Byrd's flights and discoveries. Perhaps the reason for this is that the new geographical conception of the Earth's formation in the polar regions, which is necessary to accept before the true significance of Admiral Byrd's findings can be appreciated, was not held by Navy chiefs, who, as a result, put the matter aside and forgot about it.

The above statement by Palmer that commercial airlines do not pass over the North Pole seems reasonable in the light of new Soviet discoveries in relation to the North Magnetic Pole, which was found not to be a point but a long line, which we believe is a circular line, constituting the rim of the polar concavity, so that any point on this circle could be called the North Magnetic Pole, because here the needle of the compass dips directly downward. If this is the case, then it would be impossible for airplanes to cross the North Pole, which is in the center of the polar depression and not on the Earth's surface, as according to the theory of a solid Earth and convex formation on the Pole. When pilots believe they reach the North Pole, according to compass readings, they really reach the rim of the polar concavity, where is the true North Magnetic Pole.

Referring to Giannini's book, Palmer comments:

"The strange book written by Giannini has offered the one possibility by which it can definitely be proved that the Earth is shaped strangely at the North Pole, as we believe it to be at the South Pole, not necessarily with a hole all the way through, but like a doughnut which has swelled so much in cooking that the hole is only a deep depression at each end, or like a gigantic auto tire mounted on a solid hub with recessed hub caps.

"No human being has ever flown directly over the North Pole and continued straight on. Your editor thinks it should be done and done immediately. We have the planes to do it. Your editor wants to know for sure whether such a flight would wind up in any of the countries surrounding the North Pole, necessarily exactly opposite the starting point. Navigation is not to be made by the compass or by triangulation on existing maps, but solely by gyro compass on an undeviated straight course from the moment of take-off to the moment of landing. And not only a gyro compass in a horizontal plane, but one in a vertical plane also (after one enters the polar opening). There must be a positive forward motion which cannot be disputed.

"Everyone knows that a horizontal gyro compass, such as used now, causes a plane to continually gain in elevation as the Earth curves away below it, as it progresses. Now, according to our theory of polar depression, this would mean that when a plane enters into this depression, the gyro compass should show a much greater gain in elevation than should otherwise be the case, due to the Earth's curving inward at the North Pole. Now, if the plane continues in a northerly course, this gain in altitude will continue the further it goes; and if the plane tries to maintain the same altitude, it will curve into the hollow interior of the earth."

The following statements by Giannini, written in a letter to an inquirer, who read about him in Palmer's "Flying Saucers" magazine, are interesting:

"The author was extended courtesy by the New York office of U.S. Naval Research, to transmit a radio message of godspeed to Rear Admiral Richard Evelyn Byrd, U.S.N., at his Arctic base in February, 1947.

"At that time the late Rear Admiral Byrd announced through the press, 'I'd like to see the land beyond the Pole. That land beyond the Pole is the center of the great unknown.' Subsequently, Admiral Byrd and a naval task force executed a seven hour flight of 1,700 miles over land extending beyond the theorized North Pole 'end' of the Earth.

"In January, 1947, prior to the flight, this author was enabled to sell a series of newspaper features to an international feature syndicate only because of this author's assurance to the syndicate director that *Byrd would* in fact go beyond the imaginary North Pole point.

"As a result of the author's prior knowledge of the then commonly unknown land extending beyond the pole points, and after the syndicated features had been released to the press, the author was investigated by the office of the U.S. Naval Intelligence. That Intelligence investigation was due to the fact of Byrd's definite confirmation of the author's revolutionary theories.

"Later, in March, 1958, the author delivered a radio address in Missouri, expressing the importance of the discovery of land beyond the imaginary North Pole points of archaic theory."

Speaking of the reports of Admiral Byrd's February 1947 flight beyond the North Pole, which appeared in New York newspapers, Giannini comments:

"These accounts described Byrd's 1,700 mile flight of

seven hours over *land and fresh water lakes* BEYOND the assumptive North Pole 'end' of the Earth. And the dispatches were intensified until *a strict censorship was imposed from Washington*."

Another American writer on flying saucers, Michael X, was impressed by Byrd's discoveries, and came to the conclusion that flying saucers must come from an advanced civilization in the Earth's interior, whose outer fringes Byrd visited. He describes Byrd's trip as follows:

"There was a strange valley below them. For some strange reason the valley Byrd saw was not ice-covered, as it should have been. It was green and luxuriant. There were mountains with thick forests of trees on them, and there was lush grass and underbrush. Most amazingly, a huge animal was observed moving through the underbrush. In a land of ice, snow and almost perpetual 'deep freeze,' this was a stupendous mystery.

"When Admiral Byrd went into this unknown country, into 'the center of the great unknown,' where was he? In the light of the theory of Marshall Gardner, he was at the very doorway that leads to the earth's interior and which lies *beyond the Pole.*

"Both Alaska and Canada have had much more of their share of sightings of flying saucers in recent months. Why? Is there some connection with the 'land beyond the Pole'—that unknown territory inside the earth?

"There must be a connection. If the flying saucers enter and leave the earth's interior by way of the polar openings, it is natural that they would be seen by Alaskans and Canadians much more frequently than they would be by people in other parts of the world. Alaska is close to the North Pole and so is Canada."

The above observations of a concentration of flying saucers in the Arctic region correspond to similar observations by Jarrold and Bender of a concentration in the Antarctic, where they are believed by flying saucer experts to have a landing base, from where they are seen to ascend and return. However, according to the theory of this book, what really occurs, in the Antarctic as in the Arctic, is that the flying saucers emerge from and re-enter the polar opening leading to the hollow interior of the Earth, their true place of origin. Aime Michel, in his "straight line" theory, proved that most of the flight patterns of the flying saucers are in a north-south direction, which is exactly what would be true if their origin was polar, coming from either the north or south polar opening.

In February 1947, about the time when Admiral Byrd made his great discovery of land beyond the North Pole, another remarkable discovery was made in the continent of Antarctica, the discovery of "Bunger's Oasis." This discovery was made by Lt. Commander David Bunger who was at the controls of one of six large transport planes used by Admiral Byrd for the U.S. Navy's "Operation Highjump" (1946–1947).

Bunger was flying inland from the Shackleton Ice Shelf near Queen Mary Coast of Wilkes Land. He and his crew were about four miles from the coastline where open water lies.

The land Bunger discovered was ice-free. The lakes

were of many different colors, ranging from rusty red, green to deep blue. Each of the lakes was more than three miles long. The water was warmer than the ocean, as Bunger found by landing his seaplane on one of the lakes. Each lake had a gently sloping beach.

Around the four edges of the oasis, which was roughly square in shape, Bunger saw endless and eternal white snow and ice. Two sides of the oasis rose nearly a hundred feet high, and consisted of great ice walls. The other two sides had a more gradual and gentle slope.

The existence of such an oasis in the far Antarctic, a land of perpetual ice, would indicate warmer conditions there, which would exist if the oasis was in the south polar opening, leading to the warmer interior of the earth, as was the case with the warmer territory, with land and lakes, that Admiral Byrd discovered beyond the North Pole, which was probably within the north polar opening. Otherwise one cannot explain the existence of such an oasis of unfrozen territory in the midst of the continent of Antarctica with ice miles thick. The oasis could not result from volcanic activity below the Earth's surface, for, since the land area of the oasis covered three hundred square miles, it was too big to be affected by volcanic heat supply. Warm wind currents from the Earth's interior are a better explanation.

Thus Byrd in the Arctic and Bunger in the Antarctic both made similar discoveries of warmer land areas beyond the Poles at about the same time, early in 1947. But they were not the only ones to make such a discovery. Some time ago a newspaper in Toronto, Canada, "The Globe and Mail," published a photo of a green valley taken by an aviator in the Arctic region. Evidently the aviator took the picture from the air and did not attempt to land. It was a beautiful valley and contained rolling green hills. The aviator must have gone beyond the North Pole into the same warmer territory that Admiral Byrd visited, which lies inside the polar opening. This picture was published in 1960.

In further confirmation of Admiral Byrd's discovery are reports of individuals who claimed they had entered the north polar opening, as many Arctic explorers did without knowing they did, and penetrated far enough into it to reach the Subterranean World in the hollow interior of the Earth. Dr. Nephi Cottom of Los Angeles reported that one of his patients, a man of Nordic descent, told him the following story:

"I lived near the Arctic Circle in Norway. One summer my friend and I made up our minds to take a boat trip together, and go as far as we could into the north country. So we put one month's food provisions in a small fishing boat, and with sail and also a good engine in our boat, we set to sea.

"At the end of one month we had traveled far into the north, beyond the Pole and into a strange new country. We were much astonished at the weather there. Warm, and at times at night it was almost too warm to sleep. [Arctic explorers who penetrated into the far north have made similar reports of warm weather, at times warm enough to make them shed their heavy clothing—Author.] Then we saw something so strange that we both were astonished. Ahead of the warm open sea we were on what looked like a great mountain. Into that mountain at a certain point the ocean seemed to be emptying. Mystified, we continued in that direction and found ourselves sailing into a vast canyon leading into the interior of the Earth. We kept sailing and then we saw what surprised us—a sun shining inside the earth!

"The ocean that had carried us into the hollow interior of the Earth gradually became a river. This river led, as we came to realize later, all through the inner surface of the world from one end to the other. It can take you, if you follow it long enough, from the North Pole clear through to the South Pole.

"We saw that the inner surface of the earth was divided, as the other one is, into both land and water. There is plenty of sunshine and both animal and vegetable life abounds there. We sailed further and further into this fantastic country, fantastic because everything was huge in size as compared with things on the outside. Plants are big, trees gigantic and finally we came to GIANTS.

"They were dwelling in homes and towns, just as we do on the Earth's surface. And they used a type of electrical conveyance like a mono-rail car, to transport people. It ran along the river's edge from town to town.

"Several of the inner earth inhabitants—huge giants— detected our boat on the river, and were quite amazed. They were, however, quite friendly. We were invited to dine with them in their homes, and so my companion and I separated, he going with one giant to that giant's home and I going with another giant to his home.

"My gigantic friend brought me home to his family,

and I was completely dismayed to see the huge size of all the objects in his home. The dinner table was colossal. A plate was put before me and filled with a portion of food so big it would have fed me abundantly an entire week. The giant offered me a cluster of grapes and each grape was as big as one of our peaches. I tasted one and found it far sweeter than any I had ever tasted 'outside.' In the interior of the Earth all the fruits and vegetables taste far better and more flavorsome than those we have on the outer surface of the Earth.

"We stayed with the giants for one year, enjoying their companionship as much as they enjoyed knowing us. We observed many strange and unusual things during our visit with these remarkable people, and were continually amazed at their scientific progress and inventions. All of this time they were never unfriendly to us, and we were allowed to return to our own home in the same manner in which we came—in fact, they courteously offered their protection if we should need it for the return voyage."

These giants were evidently members of the antediluvian race of Atlanteans who established residence in the Earth's interior prior to the historic deluge that submerged their Atlantic continent.

A similar experience of a visit to the hollow interior of the earth, through the polar opening, and entirely independently, was cited by another Norwegian named Olaf Jansen and recorded in the book, "The Smoky God," written by Willis George Emerson, an American writer. The book is based on a report made by Jansen to Mr. Emerson before his death, describing his real experience of visiting the interior of the earth and its inhabitants.

The title, "The Smoky God," refers to the central sun in the hollow interior of the Earth, which, being smaller and less brilliant than our sun, appears as "smoky." The book relates the true experience of a Norse father and son, who, with their small fishing boat and unbounded courage, attempted to find "the land beyond the north-wind," as they had heard of its warmth and beauty. An extraordinary windstorm carried them most of the distance, through the polar opening into the hollow interior of the Earth. They spent two years there and returned through the south polar opening. The father lost his life when an iceberg broke in two and destroyed the boat. The son was rescued and subsequently spent 24 years in prison for insanity, as a result of telling the story of his experience to incredulous people. When he was finally released, he told the story to no one. After 26 years as a fisherman, he saved enough money to come to the United States and settled in Illinois, and later in California. In his nineties, by accident, the novelist Willis George Emerson befriended him and was told the story. On the old man's death he relinquished the maps that he had made of the interior of the Earth, and the manuscript describing his experiences. He refused to show it to anyone while he was alive, due to his past experience of people disbelieving him and considering him insane if he mentioned the subject.

The book, "The Smoky God," describing Olaf Jansen's unusual trip to the hollow interior of the Earth, was published in 1908. It tells about the people who dwell inside the Earth, whom he and his father met during their visit and whose language he learned. He said that they live from 400 to 800 years and are highly advanced in science. They can transmit their thoughts from one to another by certain types of radiations and have sources of power greater than our electricity. They are the creators of the flying saucers, which are operated by this superior power, drawn from the electromagnetism of the atmosphere. They are twelve or more feet in stature. It is remarkable how this report of a visit to the Earth's interior corresponds with the other described above, yet both were entirely independent of each other. Also the gigantic size of the human beings dwelling in the Earth's interior corresponds to the great size of its animal life, as observed by Admiral Byrd, who, during his 1,700 mile flight beyond the North Pole, observed a strange animal resembling the ancient mammoth. We shall present later in this book the theory of Marshall Gardner that the mammoths found inclosed in ice, rather than being prehistoric animals, are really huge animals from the Earth's interior who were carried to the surface by rivers and frozen in the ice that was formed by the water that carried them.

Chapter II

THE HOLLOW EARTH

Before Columbus discovered America, belief in the existence of a New World across the Atlantic, in the form of a western continent, was considered as the dream of a madman.

Equally strange, in our own time, is the belief in the existence of a New World, a Subterranean World, in the hollow interior of the Earth, and which is as unknown to present humanity as the American continent was to Europeans prior to its discovery by Columbus. Yet there is no reason why it, too, may not be discovered and its existence established as a fact.

Arnoldo de Azevedo, in his "Physical Geography," wrote as follows about the mysterious world below our feet, concerning which scientists know nothing beyond a few miles in profundity, entertaining only theories, hypotheses and conjectures to hide their ignorance: "We have below our feet an immense region whose radius is 6,290 kilometers, which is *completely unknown*, challenging the conceit and competence of scientists."

This statement is absolutely true. Scientists to date have penetrated only a few miles inside the earth, and what lies further down they know nothing about, depending only on conjectures, guesses and suppositions. Many of the commonly accepted theories and beliefs about the Earth's interior do not rest on any scientific basis, and seem to originate in the old ecclesiastical idea of hellfire in the center of the Earth, which is so much like the belief of scientists that the core of the earth is a mass of fire and molten metal. Yet the scientific belief rests on no more positive evidence than the religious one. Both are merely suppositions without an iota of proof.

The belief in the Earth having a fiery center probably arose from the fact that the deeper one penetrates into the Earth, the warmer it gets. But it is a far-fetched assumption to suppose that this increase of temperature continues until the center of the Earth. There is no evidence to support this view. It is more probable that the increase of temperature continues only until we reach the level where volcanic lava and earthquakes originate, probably due to the existence of many radioactive substances there. But after we pass through this layer of maximum heat, there is no reason why it should not get cooler and cooler as we get nearer and nearer to the Earth's center.

The total surface of the Earth is 197 million square miles and its estimated weight is six sextillion tons. If the Earth was a solid sphere, its weight would be much greater. This is one among other scientific evidences of the fact that the Earth has a hollow interior.

The author believes that the truest conception of the structure of the Earth is based on the idea that when it was in a molten state during its formation, centrifugal force caused the heavier substances to be thrown outward, toward its periphery, in the form of rocks and metals, to form its outer crust, leaving its interior hollow, with openings at the poles, where centrifugal force was less and where there was less tendency to throw materials outward, which was greater at the equator, causing the bulging of the earth in this region. It has been estimated that as a result of the Earth's rotation on its axis during its formative state, polar depressions and openings thus formed would measure about 1,400 miles in diameter.

Also, we shall present below evidence to indicate that some of the original fire and incandescent materials remained in the center of the Earth to form a central sun, much smaller, of course, than our sun, but capable of emitting light and supporting plant growth. We shall also see that the Aurora Borealis or streaming lights that illuminate the Arctic sky at night come from this central sun whose rays shine through the polar opening.

Thus, if the Earth was originally a ball of fire and molten metal, some of this fire remained in its center, while centrifugal force as a result of its rotation on its axis caused its solid matter to be thrown toward the surface, forming a solid crust and leaving its interior hollow, with a fiery ball in its center, forming the central sun, which provides illumination for plant, animal and human life.

The first one to present the theory of the earth being hollow with openings at its poles was an American thinker, William Reed, author of the book, "Phantom of the Poles," published in 1906. This book provides the first compilation of scientific evidence, based on the reports of Arctic explorers, in support of the theory that the Earth is hollow with openings at its poles. Reed estimates that the crust of the Earth has a thickness of 800 miles, while its hollow interior has a diameter of

6,400 miles. Reed summarizes his revolutionary theory
as follows:

*"The earth is hollow. The Poles, so long sought, are
phantoms. There are openings at the northern and
southern extremities. In the interior are vast continents,
oceans, mountains and rivers. Vegetable and animal life
are evident in this New World, and it is probably
peopled by races unknown to dwellers on the Earth's
surface."*

Reed pointed out that the Earth is not a true sphere,
but is flattened at the Poles, or rather it begins to flatten
out as one approaches the hypothetical North and South
Poles, which really do not exist because the openings to
its hollow interior occur there. Hence the Poles are
really in midair, in the center of the polar openings and
are not on its surface as would-be discoverers of the Poles
suppose. Reed claims that the Poles cannot be discovered
because the Earth is hollow at its Pole points, which exist
in midair, due to the existence there of polar openings
leading to its interior. When explorers thought they
reached the Pole, they were misled by the eccentric behavior of the compass in high latitudes, north and south.
Reed claims that this happened in the case of Peary and
Cook, neither of whom really reached the North Pole,
as we shall see in later pages.

Starting at 70 to 75 degrees north and south latitude
the Earth starts to curve IN. The Pole is simply the outer
rim of a magnetic circle around the polar opening. The
North Magnetic Pole, once thought to be a point in the
Arctic Archipelago, has been lately shown by Soviet
Arctic explorers to be a line approximately 1000 miles

long. However, as we stated above, instead of being a straight line it is really a circular line constituting the rim of the polar opening. When an explorer reaches this rim, he has reached the ʻNorth Magnetic Pole; and though the compass will always point to it after one passes it, it is really not the North Pole even if one is deluded into thinking it is, or that he discovered the Pole due to having been misled by his compass. When one reaches this magnetic circle (the rim of the polar opening), the magnetic needle of the compass points straight down. This has been observed by many Arctic explorers who, after reaching high latitudes, near to 90 degrees, were dumbfounded by the inexplicable action of the compass and its tendency to point vertically upward. (They were then inside the polar opening and the compass pointed to the Earth's North Magnetic Pole which was along the rim of this opening.)

As the Earth turns on its axis, the motion is gyroscopic, like the spinning of a top. The outer gyroscopic pole is the magnetic circle of the rim of the polar opening. Beyond the rim the Earth flattens and slopes gradually toward its hollow interior. The true Pole is the exact center of the opening at the Poles, which, consequently, do not really exist, and those who claimed to have discovered them did not tell the truth, even if they thought they did, having been misled by the irregular action of the compass at high latitudes. For this reason, neither Cook nor Peary nor any other explorer ever reached the North or South Poles, and never will.

A very interesting article on the above subject appeared in the March 1962 issue of "Flying Saucers" magazine, written by its editor, Ray Palmer, who believes that flying saucers come from the hollow interior of the Earth through its polar openings. The article is entitled, "THE NORTH POLE—RUSSIAN STYLE." It describes remarkable discoveries made by Russian Arctic explorers, which confirm the theory of a hollow earth and polar openings, as do the observations of Arctic explorers to which we shall refer below. The article bears the following subtitle: *"More Evidence of Mystery Lands at the Poles—Two Hundred Years of Exploration Have Given the Russians a New Concept of the Pole and Render all Previous Geographies Obsolete—Here are Indisputable Geophysical Facts!"*

We shall now quote from this article:

"Many readers will remember the articles we have published giving our theories that there is something mysterious about each polar area of the Earth. We have suggested that there is much more 'area' at both poles than it is possible to show on a globe map. We have pointed out Admiral Byrd's strange flights 'beyond' the pole. We have mentioned the case of missing mountains and different branches of the military discounting the mapping ability of the other. We have even suggested that the Earth is hollow, and that giant 2,100 mile openings exist at the poles, and there is much evidence of the existence of these openings. We have pointed out that there is a great deal of secrecy and double-talk about the Arctic and Antarctic areas. We have even suggested that the flying saucers might come from this mystery area, or from inside the Earth.

"One of the things we have been most insistent about

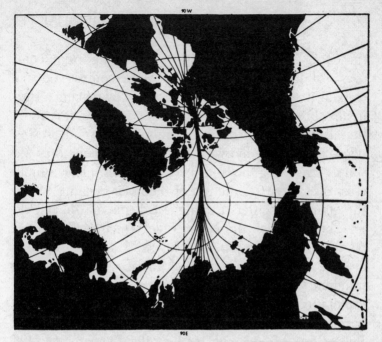

THE NORTH MAGNETIC POLE, once thought to be virtually a point in the Arctic Archipelago, has been shown by recent investigations to extend across the polar Basin to the Taimyr Peninsula in Siberia. The lines represent magnetic meridians. (*Latest Scientific Conception of the North Magnetic Pole: Based on Researches by Russian Scientists.*)

is that no one has yet been to the North Pole, all claims to having done so being false, because the Pole is not a 'point,' and cannot be 'reached' in the accepted sense of the word.

"We have successfully challenged those military and civilian pilots who have claimed that they fly 'daily' over the North Pole. In the case of the military flyer we have pointed out the maneuver which is standard, which automatically makes it impossible for him to fly 'beyond' the Pole by flying straight across it. [That is, across the polar opening, instead of going into it—Author.] Be-

cause of 'navigating difficulties stemming from compasses of all kinds,' a 'lost' flier (whose compass doesn't
work as it should) regains his bearings by making a turn
in any direction, until his compasses again resume func-
tion. In the case of commercial airlines, whose advertising
boast is that they fly twice daily over the Pole, they are
simply stretching the truth by 2,300 miles. [They simply
cross over the magnetic rim of the polar opening, where
the compass registers the highest degree north, but do
not actually reach the North Pole, which is the central
point of the polar opening inside this rim—Author.]

"We have available, in the form of records of several
hundred years, in Russian archives, a history of Arctic
exploration which proves our most important point
beyond further question: i.e., that the North Magnetic
Pole is not a point, but (deduce the Russians) a 'line'
approximately 1000 miles long. Before we go further,
we might suggest that we think they are wrong in this
deduction, and that instead of being a line, it is actually
a circle. Because of lack of space to place it on the
globe, the Russians have been forced to compress their
observations into a two dimensional area. They had to
squeeze the circle from two sides and make a line out of
it. We'd like to give you now a resume of that single
point of Russian exploration, which actually covers much
more than just geomagnetism.

"Here is what the Russians say: Navigators in the
high latitudes have always been troubled by the odd
behavior of their magnetic compasses caused by apparent
irregularities and asymmetries in the magnetic field of

the Earth. Early magnetic maps have been drawn on this
assumption, based on hopeful guesses, that the North
Magnetic Pole is virtually a point. Accordingly, it was
expected that the compass needle, which dips more
steeply as it approaches the Magnetic Pole, would point
straight down, or very nearly so, at the Magnetic Pole
itself. But data from many Russian and other expeditions
showed that the compass needle points straight down for
a very long distance across the Arctic Ocean, from a point
northwest of the Taimyr Peninsula to another point in
the Arctic Archipelago. This discovery first inspired the
hypothesis that there is a second North Magnetic Pole,
tentatively located at 86 degrees East longitude. More
refined observation has disposed of this idea. The map
of the magnetic field now shows the magnetic meridians
running close together in a thick bunch of lines from the
North Magnetic Pole in the Arctic Archipelago to
Siberia. *The North Magnetic Pole, once thought to be
virtually a point in the Arctic Archipelago, has been
shown by recent investigations to extend across the
polar basin to the Taimyr Peninsula in Siberia.*

"The 'Pole,' magnetically speaking, is a very extended
area that crosses the Polar Basis from one continent to
the other. It is at least 1,000 miles long, and more likely
can be said to exist as a rather diffused line for 1,000
miles more. [It is really not a point in the far north, but
is the rim of the polar opening, since after Admiral Byrd
passed it and entered the polar opening leading to the
Earth's interior, he left the Arctic ice and snow behind

and entered a warmer territory—Author.] Thus when Admiral Peary (and any other Arctic explorer who used a magnetic compass) claims to have 'reached' the Pole, he is making a very vague claim indeed. He can only say that he reached a point, which can be anywhere in a demonstrable 2,000 mile area (the magnetic rim of the polar opening), where his compass pointed straight down. A noteworthy achievement, but not a 'discovery of the Pole.'

"Since other types of compass, such as the gyroscopic and the inertial guidance, have equally vague limitations, we make bold to say that *nobody ever reached the Pole*, and more, there is not a 'Pole' to reach.

"Next, having found themselves stumped to account for the strange behavior of the compass in the Polar Basin, the theorists have turned to space and the upper atmosphere and even to the sun for an explanation of what is happening to their instruments. Now the Pole has become 'the interaction of the magnetic field with charged particles from the sun.'

"More significant are the unfavorable references to former cartographers whose maps are now 'thick clouds congealed in the imagination of cartographers as land masses.' The Navy, as an example, feels a bit put out when the Army says their missing South Pole mountains were never there, because the Army cannot find them by their own confused reckoning based on a magnetic pole which 'isn't there at all.' We find now that new land areas are 'discovered' and old maps tossed out because the lands they show are not there any more. [This confusion is due to the irregular action of the compass in the far north due to the fact that the North Magnetic Pole is not a point as former cartographers supposed, but a circle around the rim of the polar opening—Author.]

"This brings us to the subject of 'mystery lands' of great extent in the polar areas, which cannot possibly be placed on our globe without overlapping seriously in impossible ways. . . . Could it be here where the flying saucers originate?"

It is well known that the North and South Magnetic Poles do not coincide with the geographical poles, as they *should* were the Earth a solid sphere, convex at its poles. The reason why the magnetic and geographical poles don't coincide is because, while the magnetic pole lies along the rim of the polar opening, the geographical pole lies in its center, in midair and not on solid land. As we shall see below, the true magnetic pole is not on the external rim of the polar opening but the center of the Earth's crust, which should be about 400 miles below the surface, and running around the polar opening. For this reason the needle of the compass still continues to point vertically downward after one passes the rim of the polar opening and penetrates into it. Only after passing its center would the needle of the compass start pointing upward instead of downward, but in either case, after reaching the rim of the polar opening, the compass no longer functions horizontally, as previously, but vertically. This has been observed by all Arctic explorers who reached high latitudes and puzzled them. The only explanation is provided on the conception of a hollow earth and polar openings, with the magnetic pole and center of gravity in the middle of the Earth's crust, and not in its geometrical center. As a result, ocean water on the inside of the crust adheres to its inner surface just as it does on the outside. We may calculate the Earth's magnetic pole and center of gravity as a circular line around the polar opening, but in its middle, about 400 miles from the Earth's surface.

In support of the above conception regarding the magnetic pole being situated in the rim of the polar opening, Palmer refers to the following facts: Between each magnetic pole around the Earth pass magnetic meridians. In contrast with geographical meridians, which measure longitude, the magnetic meridians move from east to west and back again. The difference between the geographical meridians, or true north and south, and the direction in which a magnetic compass points, or the magnetic meridian of the place, is called the declination. The first observation made was in London in 1580 and showed an easterly declination of 11 degrees. In 1815 the declination reached 24.3 degrees westerly maximum. This makes a difference of 35.3 degrees change in 235 years, which is equal to 2,118 miles. Now if we make a circle around the Pole, with a radius of 1,059 miles, so that is it 2,118 miles in diameter, this would represent the rim of the polar opening along which, in this case, the North Magnetic Pole traveled from one point to its diametrically opposite point on the circle, 2,118 miles away, in 235 years. This is the reason why the magnetic pole and the geographical pole do not coincide. The geographical pole is an extension of the Earth's axis and since this runs through the center of the polar opening, it exists in empty space—hence can never be "discovered" by any explorer, since it is not on solid land.

According to Marshall Gardner, the rim of the polar opening, which is the true magnetic pole, is a large circle 1,400 miles in diameter. It is so large that when explorers pass it, as many did, the slope is so gradual that they never know they are entering the interior of the Earth, but imagine they are on the surface. The magnetic pole can therefore be any point on the circle of the magnetic rim of the polar opening. On this point, Palmer says:

"The focal point, or the actual 'pinpoint' of the magnetic pole exists on only one portion of the circumference of that circle at a time, and moves progressively around the circle in a definite 'orbit' that takes some 235 years. This would make the magnetic pole travel approximately 18 miles per year.

"Military and civilian flights 'over the Pole' can be made daily without producing the slightest evidence of the vast hole in the Earth, whose perimeter they circumscribe, no matter what they ASSUME in their navigational procedure, due to the original error in assumption that what they are passing over is a POINT and not a vast CIRCUMFERENCE which they touch at only one place, and then immediately deviate away from its natural curve because they are traveling in a straight line."

If the Earth was a solid sphere, with two poles at the end of its axis, being a magnet, its magnetic poles would coincide with its geographical poles. The fact that they do not is inexplicable on the basis of the theory that it is a solid sphere. The explanation becomes clear when we assume the existence of polar openings, with magnetic poles along the circular rim of these openings, rather than at a fixed point.

Palmer quotes a significant statement by Russian Arctic explorers who say: "Exploration and research have shown that an enormous area of the Earth's surface and correspondingly *large realms of the unknown* may be brought within the compass of human understanding in a very few years."

This statement by the Russians sounds remarkably similar to Admiral Byrd's statements about the trans-Arctic region being "the center of the Great Unknown." Could it be that the Russians know about Admiral Byrd's discovery of "a vast new territory" beyond the Pole? Palmer comments on this Russian statement as follows:

"This is truly a stupendous sentence. Contemplate what it actually says. It says that not only exploration, but also 'research' have shown that enormous regions of the Earth's surface AND correspondingly (this word is significant) large realms of the UNKNOWN may be brought within the compass of UNDERSTANDING of human beings in a very few years. In plain words, in addition to areas we can understand and investigate by exploration, there are large realms which have to be brought to human understanding by means of research.

"Yes, large UNKNOWN and even BEYOND PRESENT UNDERSTANDABILITY areas do exist, and it 'MAY BE' that we will discover and comprehend them in a very few years. In plain words, in addition to areas we can understand and investigate by exploration, there are large realms which have to be brought to human understanding by means of research.

"In the next few sentences (of the Russians) we find that there is much 'prospect for development' in a Polar Basin which, by present concepts, is nothing but frozen ocean. What is it that is such a great prospect for development? Ice cubes for our tea? No, there must be very much more interesting possibilities, the kind of possibilities that entail large land masses in an *unknown area* yet to be explored and developed."

Palmer quotes the Russians as saying: "As recently as 30 years ago more than half the total area of the Polar Basin was unexplored, and 16 percent was still *terra incognita* only 15 years ago. Today, disappointing as this may be to young geographers, the area of blank spots on the map of the Polar Basin has shrunk to almost nothing. At the same time, to the regret of the older explorers and the understandable pleasure of the younger ones, there are still blank spots elsewhere in the Arctic. The ocean, the air and the ionosphere still hold many mysteries."

Palmer comments on this Russian statement: "We learn that the blank spots on the map of the Polar Basin have shrunk to almost nothing. In the next breath we find that there are still blank spots ELSEWHERE in the Arctic. Where else? The ocean, the air and the ionosphere, they say, still hold many mysteries. Particularly the ocean, in the UNKNOWN extent of which exist vast land masses so far not only beyond our ability to place on our maps, but beyond our ability to understand.

"We might say all this is double talk. We might also say secrets are being kept. But we won't. The fact is that neither is true. It is STRAIGHT talk, the only kind of talk we can expect from anyone who is trying to tell something, but cannot because it is, as yet, beyond his understanding. To say definitely that there are large land masses inside an area commonly called a 'point' is to be faced with a challenge to demonstrate and prove. Since this cannot be done, the speaker is left rather helpless to do more than hint vaguely at mysteries.

"It is up to the opponents of the 'Mystery Land at the Pole' theory to disprove it, or prove their own—and their own has been irrevocably demolished by the scientists and explorers of the two greatest nations on earth. What we have presented is not a theory—but the cumulative result of hundreds of years of exploration, culminated by the geographical year which established the information we have given you as the 'new concept of geomagnetism in the Polar Basin.'

"The mystery is at last coming to the fore, and the scoffers are at last silenced. Let us all work together to dig out the truth about this mystery that is so engrossing, and so important to mankind. What is it that exists at both Poles of the earth, which opens to us new frontiers so vast in extent and nature as to be beyond present understanding? It may well be that exploration of space is far less important than the exploration of our mysterious planet, which has now suddenly become a 'vast realm' far larger than we ever dreamed it to be."

The theory of a hollow earth with openings at the poles was originated by William Reed in 1906, when he first presented it in his book, "Phantom of the Poles." Fourteen years later, in 1920, another American writer,

Marshall B. Gardner, published a book entitled "A Journey to the Earth's Interior or Have The Poles Really Been Discovered?" Apparently he knew nothing about Reed's book, since he did not mention it in his bibliography, which was quite extensive and included most of the important books on Arctic exploration, which he quoted in support of his theory of a hollow earth.

Gardner, in his book, presents the same conception of the Earth's structure as Reed did, claiming that it is hollow, with openings at its Poles, but he differs from Reed in that he believes in the existence of a central sun which is the source of the aurora borealis. In the diagrams of his book, Gardner depicts the Earth as having circular openings at its poles; and the ocean water, which flows through these openings, adheres to the solid crust, both above and below, since the center of gravity of the Earth, according to his theory, resides in the middle of this solid portion and not in its hollow interior. For this reason, if a ship travels through the polar opening and reaches the Earth's interior, it would continue to sail in a reversed position on the inside of the crust, just as, at night, we are below the Earth's surface held to it by gravity.

Gardner's book, which is now out of print and very rare, seeming to have met the fate of other writings on this subject by being lost and forgotten and its message unknown to the world at present, has many interesting diagrams, some of which we are reproducing. We quote his description of these diagrams:

1. "Showing the Earth bisected centrally through the polar openings and at right angles to the Equator, giving

a clear view of the central sun and interior continents and oceans (Reproduced from a working model, made by the author in 1912.)

2. "The Earth as it would appear if viewed from space, showing the north polar opening to the Earth's interior, which is hollow and contains a central sun instead of an ocean of liquid lava."

3. "Diagram showing the Earth as a hollow sphere with its polar openings and central sun. The letters at the top and bottom of the diagram indicate the various steps of an imaginary journey through the planet's interior. At the point marked 'D' we catch our first glimpse of the corona of the central sun. At the point marked 'E' we can see the central sun in its entirety."

Gravitational pull is strongest around the curve from the exterior to the interior of the Earth. A 150 pound man would probably weigh 300 pounds while sailing through the polar opening and around the curve from the outside to the inside of the Earth. When he reached the inside he would weigh only 75 pounds. This is because less force is needed to hold a body to the inside of a hollow ball in rotation than to hold it to the outside, due to centrifugal force.

William Reed says that gravitational pull is strongest about half way around the curve leading to the interior of the Earth, where is the center of gravity, being so strong there that the salt water and fresh water of icebergs (which, as we shall see below, come from the Earth's interior) do not mix. The salt water remains a few feet below the fresh water. This enables one to obtain fresh drinking water from the Arctic Ocean. But how can fresh water be found in the extreme north, where there is only salty ocean water, and how can icebergs be formed of fresh water, not salt water? The only explanation, as both Reed and Gardner point out, and as we shall see below, is that this fresh water comes from rivers that arose in the Earth's warmer interior, which, after they reach the colder surface, suddenly freeze and turn into icebergs, which break off and fall into the sea, producing the strange tidal waves that Arctic explorers have observed in the far north, and which puzzled them.

Both Reed and Gardner claim that the temperature in the inside of the Earth is much more uniform than on the outside, being warmer in winter and cooler in summer. There is adequate rainfall, more than on the surface, but it is never cold enough to snow. It is an ideal subtropical climate, which is free from the oppressive heat of the tropics, as well as from the cold weather of the temperate zone. They also claim that the north polar opening is larger than the south. They say that there exists a Land of Paradise on the other side of the Mammoth Ice Barrier, which must be passed before one reaches a warmer climate in the land that lies beyond the Pole, over which Admiral Byrd flew.

Around the curve at the polar opening is another ring of ice, called the Great Massive Fresh Water Ice Pack or Ice Barrier. Here is where icebergs originate. Each winter this ring of ice is formed from fresh water which flows out from the *inside* of the Earth. During the winter months, billions of tons of free-flowing fresh water, com-

The earth as it would appear if viewed from space showing the north polar opening to the planet's interior which is hollow and contains a central sun instead of an ocean of liquid lava. (*Reproduced from "A Journey To The Earth's Interior—or—Have The Poles Really Been Discovered," by Marshall B. Gardner. Printed by Eugene Smith Company, Aurora, Illinois, 1920.*)

ing from rivers inside the Earth and flowing toward the outside through the polar openings, freeze at their mouth and form mountains of fresh water ice, whose presence in this region would be inexplicable if the Earth was a solid sphere. In summer time, huge icebergs, miles long, break off and float to the outside of the Earth. They are composed of *fresh* water, when there could exist only salt water at the poles. Since this is the case and since all

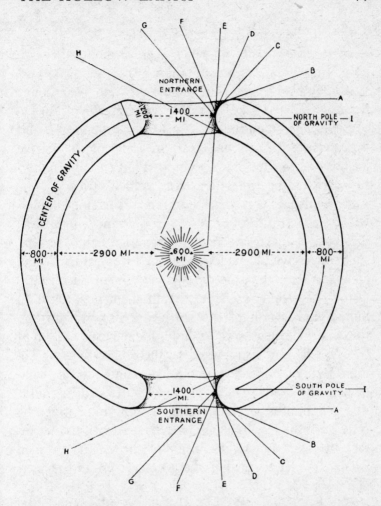

Diagram showing the earth as a hollow sphere with its polar openings and central sun. The letters at top and bottom of diagram indicate the various steps of an imaginary journey through the planet's interior. At the point marked "D" we catch our first glimpse of the corona of the central sun; at the point marked "E" we can see the central sun in its entirety. (*Reproduced from "A Journey To The Earth's Interior—or—Have The Poles Really Been Discovered," by Marshall B. Gardner. Printed by Eugene Smith Company, Aurora, Illinois, 1920.*)

water on the outside of the Earth in these regions is salty, the fresh water of which these icebergs are composed must come from its interior.

Inside the icebergs, the mammoth and other huge tropical animals, believed to be of prehistoric origin because never seen on the Earth's surface, have been found in a perfect state of preservation. Some of them have been found to have green vegetation in their mouths and stomachs at the time they were suddenly frozen. The usual explanation is that these are prehistoric animals which lived in the Arctic region at the time when it had a tropical climate, and that the coming of the Ice Age, suddenly converted the Arctic from a tropical to a frigid zone and froze them before they had time to flee southward. The great ivory deposits from elephants, found in Siberia and islands of the north, are also explained in this way. Gardner, however, holds to an entirely different theory, which was supported by the observations of Admiral Byrd of a huge mammoth-like creature in the "land beyond the Pole," which he discovered. Gardner claims that mammoths are really animals now inhabiting the interior of the Earth, which have been carried to the surface by rivers and frozen inside of the ice that formed when the rivers reached the surface, forming glaciers and icebergs.

In Siberia, along the Lena River, there lie exposed on the soil and buried within it, the bones and tusks of millions of mammoths and mastodons. The consensus of scientific opinion is that they are prehistoric remains, and that the mammoth existed some 20,000 years ago, but was wiped out in the unknown catastrophe we now call the last Ice Age.

It was Schumachoff, a fisherman living in Tongoose, Siberia, who, in 1799, first discovered a complete mammoth frozen in a clear block of ice. Hacking it free, he removed its huge tusks and left the carcass of fresh meat to be devoured by wolves. Later an expedition was sent to examine it, and today its skeleton may be seen in the Museum of Natural History in Leningrad.

Polar explorers not only mention fauna (animals) but flora (vegetation) in the extreme north. Also many animals, like the musk-ox, strangely migrate northward in winter, which it would do only if it reached a warmer land there. Repeatedly, Arctic explorers have observed bears heading northward into an area where there could not be food for them if there was no polar opening into a warmer region. Foxes also were found north of the 80th parallel heading north, obviously well fed. Without exception, Arctic explorers agree that, strangely, *the further north one goes, after a certain latitude, the warmer it gets*. Invariably, a north wind brings warmer weather. Coniferous trees were found drifting ashore, coming from the far north. Butterflies and bees were found in the far north, and even mosquitoes, but they are not found hundreds of miles to the south and not until Canadian and Alaskan climate areas conducive to such insect life are reached.

Unknown varieties of flowers were also found in the extreme north. Birds resembling snipe, but unlike any known species of bird, were seen to come from the north, and to return there. Hare are plentiful in a far northern area where no vegetation grows but where vegetable matter is found in drifting debris from the more northern open waters.

Eskimo tribes have left unmistakable traces of their migration by their temporary camps, always advancing northward. Southern Eskimos speak of tribes that live in the far north. They hold the belief that their ancestors came from a land of paradise in the extreme north.

In New Zealand and lower South America are found identical fauna and flora which could not have migrated from one of these places to the other. The only explanation is that they came from a common motherland— the Antarctic continent. Yet how could they come from there if it is a frozen waste where only penguins seem able to survive? "Only Admiral Byrd's 'mystery land' can account for these inexplicable facts and migrations," concludes Palmer.

Many Arctic explorers, after passing the ring of ice around the curve leading to the Earth's interior, continued straight north until they crossed this ice barrier. Many entered the opening leading to the interior but did not know it and thought they were still on the outer surface. The reason for this is that the opening is so large that one cannot know the difference except that the sun rises later and sets sooner, its rays being cut off by the rim of the polar opening after one enters it. This has been observed by all Arctic explorers who went sufficiently north. The polar opening is believed by Gardner to be 1,400 miles in diameter.

Once they were inside the Earth, explorers entered a

New World where they found things opposite to what they expected. The needle of the compass pointed vertically instead of horizontally as it did before, due to the fact that the true magnetic pole is located in the middle of the curve leading from the outside to the inside of the earth. The further north they went, the warmer it became. The ice of Arctic regions further south disappeared and was replaced by open sea. (Admiral Byrd found a total absence of ice and snow in the "land beyond the Pole" over which he passed for 1,700 miles.) As explorers sailed further north, the north winds became warmer and warmer. The weather was mild and pleasant. Often the dust, carried by the wind, was unbearable. Some explorers, like Nansen, had to turn back due to the dust. Where could this dust come from in the extreme north, a land of ice and ocean? Reed and Gardner ascribe the origin of this dust, often noticed by Arctic explorers, to volcanoes inside the polar opening leading to the interior of the Earth. It would be impossible to expect volcanoes in the Arctic, except if they were inside the polar opening.

On August 3, 1894, Dr. Fridtjof Nansen, an Arctic explorer, in the far north, was surprised at the warm weather there and the fox tracks he found. He was probably inside the polar opening then. His compass utterly failed to work, so that he did not know where he was. The further into the opening he went, the warmer it became. If he had gone still further he would have seen tropical birds, as other explorers did, as well as other animals not seen on the Earth's surface, as the mammoth that Admiral Byrd observed when he looked down from his plane, during his 1,700 mile flight over this mysterious icefree Arctic area.

Ray Palmer writes:

"The musk-ox, contrary to expectations, migrates north in the wintertime. Repeatedly, Arctic explorers have observed bears heading north into an area where there cannot be food for them. Foxes also are found north of the 80th parallel, heading north, obviously well fed. Without exception, Arctic explorers agree that the further north one goes, the warmer it gets. Invariably a north wind brings warmer weather. Coniferous trees drift ashore from out of the north. Butterflies and bees are found in the far north, but never hundreds of miles further south; not until Canadian and Alaskan climate areas conducive to such insect life are reached.

"Unknown varieties of flowers are found. Birds resembling snipe, but unlike any known species of bird, come out of the north, and return there. Hare are plentiful in an area where no vegetation ever grows, but where vegetation appears as drifting debris from the northern open water. Eskimo tribes, migrating northward, have left unmistakable traces of their migration in their temporary camps, always advancing northward. Southern Eskimos themselves speak of tribes that live in the far north. The Ross gull, common at Point Barrow, migrates in October toward the North. Only Admiral Byrd's 'mystery land' can account for these inexplicable facts and migrations."

The Scandinavian legend of a land of paradise in the far north, known as "Ultima Thule," commonly confused with Greenland, is significant because, centuries before Admiral Byrd's flight, the existence of such an icefree land in the northern limits of the Earth was anticipated. Palmer writes:

"The Scandinavian legend of a wonderful land far to the north called 'Ultima Thule' (commonly confused with Greenland) is significant when studied in detail, because of its remarkable resemblance to the kind of land seen by Byrd, and its remarkable far north location. To assume that Ultima Thule is Greenland is to come face to face with the contradiction of the Greenland Ice Cap, which fills the entire Greenland basin to the depth of 10,000 feet. Is Admiral Byrd's land of mystery, the center of the great unknown, the same as the Ultima Thule of the Scandinavian legends?

"There are mysteries concerning the Antarctic also. Perhaps the greatest is the highly technical one of biology itself; for on the New Zealand and South American land masses are identical fauna and flora which could not have migrated from one to the other, but rather are believed to have come from a common motherland. That motherland is believed to be the Antarctic Continent. But on a more popular level is the case of the sailing vessel 'Gladys,' captained by F. B. Hatfield in 1893. The ship was completely surrounded by icebergs at 43 degrees south and 33 degrees west. At this latitude an iceberg was observed which bore a large quantity of sand and earth, and which revealed a beaten track, a place of refuge formed in a sheltered nook, and the bodies of five dead men who lay on different parts of the berg. Bad weather prevented any attempts at further investigation.

"An unanimous consensus of opinion among scientists

is that one thing peculiar to the Antarctic is that there are no human tribes living upon it. Also investigation showed that no vessel was lost in the Antarctic at the time, so that these men could not be shipwrecked sailors. Could it be that these men who died on the berg came from 'that mysterious land beyond the South Pole' discovered by the Byrd expedition? Had they ventured out of their warm habitable land and lost their way along the ice shelf, finally to be drifted to their deaths at sea on a portion of it, broken away to become an iceberg while they were on it?"

Another American writer on the subject of the Earth being hollow, named Theodore Fitch, referring to the ice barriers that must be crossed before one can enter the polar openings leading to the Earth's interior, asks: "Why can't we fly over these huge ice barriers or make roads and travel overland over them to the inside of the Earth?" He sees no reason why this cannot be done, even though he, like most other Americans, was in total ignorance of the fact that Admiral Byrd flew over these ice barriers some years before, and had entered this new territory. Fitch believes that once these facts are made public, every large nation will try to establish a foothold in this New World, whose land area is greater than that on the Earth's surface and which is free from radioactive fallout to poison its soil and foods. This New World could more easily be reached than the moon and is of much more importance to us, since it provides ideal conditions for human life, with a better climate than exists on the surface. Fitch calls it a Land of Paradise, and believes it is the true geographical location of Paradise, a wonderful land referred to in the religious writings of all peoples.

It seems that the Russians are now doing what Fitch suggested by sending fleets of icebreakers, some atomic-powered, to explore the far north. The next step will be for the Russians to repeat Admiral Byrd's flight through the polar opening to the "land beyond the Pole."

Fitch's book is entitled "Our Paradise Inside the Earth." He based it on the works of Reed and Gardner. He mentioned that during the last century a sea captain, who traveled due north, curved inward into the interior of the Earth, though he thought he was heading toward the North Pole. Fitch writes:

"Both William Reed and Marshall Gardner declare that there must be a land of paradise on the other side of the mammoth ice barrier. Both men are of the opinion that a race of little brown people live in the interior of the Earth. It is possible that the Eskimos descended from these people.

"Most explorers have sailed straight north until they went around the 800-mile curve at the polar opening. Not one of them knew they were on the inside of the Earth. These explorers found things exactly opposite from what they expected. As they sailed north, the north winds became warmer and warmer. Except for strong dusty warm winds once in a while, the weather was mild and pleasant. Except for icebergs from the interior, the sea was open and sailing good. [Reed and Gardner explain this strange dust found in the very far north and which darkens the snow on which it falls, as we have pointed out above, as coming from active volcanoes inside the polar opening. This seems to be the only possible explanation—Author.]

"They saw countless square miles of good land. The further north they went, the more grass, flowers, bushes, trees and other green vegetation they saw. One explorer wrote that his men gathered eight different kinds of flowers. They also reported that they saw sloping hills covered with green vegetation. [These observations were confirmed by Admiral Byrd, who, during his 1,700-mile flight over this iceless territory, saw trees, vegetation, mountains, lakes and animal life—Author.]

Another writer said he saw all kinds of warm-weather animals and millions of tropical birds. They were so thick that a blind man could bring down one or more birds with one shot. The lovely scenery of both sky and land was more magnificent than anything ever seen on the exterior of the Earth. Each explorer wrote about the majesty of the aurora borealis or Northern Lights. It is claimed that the Northern Lights really result from the light of the central sun inside the Earth shining through the opening at the North Pole."

Fitch points out that the hollow interior of the earth has a land area larger than the outer surface because while 75 per cent of the earth's surface is covered with water, leaving only 55 million square miles of land surface, the total surface of the earth is 197 million square miles. Fitch claims that there are no oceans in the interior comparable in size with those on the surface, and that there are three times as much land inside the earth as on the outside so that in spite of the smaller circumference and less total area of the interior, its land area is greater. Fitch says that it has a better and healthier climate than we have on the surface, without cold winters, hurricanes, earthquakes, electric storms, cyclones, radioactive fallout, nefarious cosmic rays, radioactive solar radiations, soil erosion from excessive rainfall and other disadvantages. It has an ideal subtropical climate.

Another American writer who was much influenced by the theories of Reed and Gardner is William L. Blessing, who published a booklet on the subject in which he reproduced their diagrams of the Earth's structure. Blessing wrote:

"The Earth is not a true sphere. It is flat at the poles, or, I should say, it begins to flatten out at the poles. The pole is simply the outer rim of a magnetic circle, and at this point the magnetic needle of the compass points down. As the earth turns on its axis, the motion is gyroscopic. The outer gyroscopic pole is the magnetic rim of a circle. Beyond the rim the Earth flattens and slopes gradually like a canyon into the interior. The true pole in the exact center of the cone is perpendicular, for this point is the exact center of the opening or hollow into the Earth's interior.

"The old idea that the Earth was once a solid or molten mass and that the center is composed of molten iron must be discarded. Since the shell of the Earth is about 800 miles thick, that would mean that the molten iron core would be more than 7,000 miles in diameter and 21,000 miles in circumference. Impossible.

"Likewise, the old idea that the deeper into the Earth the hotter it becomes must also be discarded. It is radium and radioactivity that produce the heat in the earth.

All surface rocks contain minute particles of radium."

One of the most puzzling facts of Arctic exploration is that while the area is oceanic, covered with water, which is variously frozen over or partially open, depending on the time of the year, many explorers have pointed out, paradoxically, that the open water exists in greater measure at the points nearest to the Pole, while further south there is more ice. In fact, some explorers found it very hot going at times, and were forced to shed their Arctic clothing. There is even one record of an encounter with naked Eskimos. In fact, the origin of the Eskimo race is believed to be in the extreme north, from where they migrated southward to their present habitat. Their original more northern home was probably warmer than their present more southern one.

It is strange that Reed's and Gardner's books, which presented such an epoch-making geographical theory, which they supported by the evidence of Arctic exploration during the past century—a theory comparable in importance to the theory that the Earth is round, when it was first proposed—should have been so disregarded (or were they suppressed?), so that today they are unavailable and very rare. (It was the author's good fortune to secure a copy of Gardner's book from a bookdealer handling rare books.) Is it possible that these books shared the fate of the news about Admiral Byrd's discoveries, Giannini's book and Palmer's magazine announcing Byrd's confirmation of Reed's and Gardner's theory of a hollow Earth with openings at the poles? (A correspondent of the author's, living in Washington, D.C., wrote that he happened to look through the books in the library of a high official of the Air Force, with whom he had business, and, much to his surprise, he saw a copy of Gardner's book.) Evidently Gardner's theory of a hollow Earth is not unknown to government and military leaders in view of Admiral Byrd's having confirmed it; but it is hushed up and not openly discussed.

Fitch asks those who do not believe that the Earth is hollow, with openings at its poles, to answer the following questions:

"Can you produce proof that any explorer reached the so-called North or South Pole?

"If there is no such thing as 83 to 90 degrees latitude ON the Earth, then how can one reach or fly over the North Pole?

"If the Earth is not hollow, then why does the north wind in the Arctic get warmer as one sails north beyond 70 degrees latitude?

"Why are there warm northerly winds and an open sea for hundreds of miles north of 82 degrees latitude?

"After 82 degrees latitude is reached, why is the needle of a compass always agitated, restless and balky?

"If the Earth is not hollow, then why do the warm northerly winds mentioned above carry more dust than any wind on earth?

"If no rivers are flowing from the inside to the outside, then why are all icebergs composed of fresh water?

"Why does one find tropical seeds, plants and trees floating in the fresh water of these icebergs?

"If not all the fresh water icebergs positively do not come from any place ON earth, as would be impossible unless we assume the existence of rivers flowing from the inside to the outside, then where do they come from?

"If the inside of the Earth is not warm, why do millions of tropical birds and animals go further north in the winter time?

"Why does the wind from the north carry more pollen and blossoms than any wind on the exterior?

"If it is not hollow and warm inside the Earth, then why does colored pollen color the snow for thousands of square miles?

"Could it be that pollen from millions of acres of colored flowers causes the snow to be red, pink, yellow, blue, etc.?"

Showing the earth bisected centrally through the polar openings and at right angles to the equator, giving a clear view of the central sun and the interior continents and oceans. (*Reproduced from photograph of working model. Made by the author, 1912. Patented May 12, 1914, No. 1096102.*)

Chapter III

WILLIAM REED'S BOOK,
"PHANTOM OF THE POLES"

Presenting Scientific Evidence, Based on Arctic
Exploration, to Prove for the First Time that the
Earth is Hollow With Openings at the Poles

In 1906 appeared the first book to offer scientific proof
that old geographical conceptions about the earth's
structure are false and that the earth, instead of being a
solid sphere, as commonly assumed, is really hollow, with
openings at the poles. Were this a book created from the
author's imagination, it might be disregarded as a work
of science fiction—but since the book is based on an extensive bibliography representing the reports of Arctic explorers, it must be taken more seriously.

This book was published in New York and written by
William Reed. Its title was "The Phantom of the Poles,"
and claimed the Poles were never discovered because
they do not exist. Where the North and South Poles are
supposed to be located, Reed claims, are huge polar
openings in which the Poles are in the center, for which
reason they can never be reached by any explorer.

Reed's book was written fourteen years before that of
Marshall Gardner, who claimed that not only was the
earth hollow but that there was a central sun at its center.
Reed, however, did not include this central sun in his
theory, but believed that the higher temperature in the

region of the Poles is due to burning volcanoes at the polar openings, which are the origin of the dust that Arctic explorers noticed there. We now quote from Reed's book. On page 282 he says:

"The earth is either hollow or it is not. What proof have we that it is not hollow? None at all that is positive and circumstantial. On the contrary, everything points to its being hollow. If it be so, and if there are burning volcanoes in the interior, would you not see great lights reflected on the icebergs and clouds, just as other great fires reflect the light? Would not great clouds of smoke and dust be seen—the same as from any other burning volcano? That is what all the explorers have witnessed— low dark clouds rising from the ocean, or at the edge of the ice. Nansen (an Arctic explorer) said: 'Let us go home! What have we here to stay for? Nothing but dust, dust, dust!'

"Where could such dust come from—so bad that it was one of the great annoyances in the heart of the Arctic Ocean, if it did not come from an exploding, burning volcano (in the polar opening)?

"If the earth be hollow, would it not be warmer in winter and cooler in summer (as we enter the polar opening)? Arctic explorers say that a north wind in winter raises the temperature, while a south wind lowers it. As an opposite fact, in summer a south wind raises the temperature, while a north wind lowers it. That is just what would occur if the winds come from the interior of the earth. Again, if the earth is hollow, it could not be round, inasmuch as the opening would take from its roundness in proportion to the size of the opening. All now agree that the earth flattens at the poles. Also it is warmer the further one goes north or south. Why is this the case?

"There is but one answer, and that is that the earth is

GLOBE SHOWING SECTION OF THE
EARTH'S INTERIOR

The earth is hollow. The poles so long sought are but phantoms. There are openings at the northern and southern extremities. In the interior are vast continents, oceans, mountains and rivers. Vegetable and animal life are evident in this new world, and it is probably peopled by races yet unknown to the dwellers upon the earth's exterior. THE AUTHOR. (*Reproduced from "The Phantom Of The Poles" by William Reed, published by Walter S. Rockey Company, New York, 1906.*)

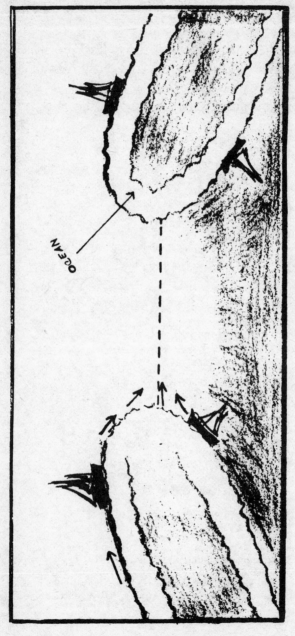

OCEAN

SHIPS SAILING TO THE INSIDE

The Bible, the Book of Enoch, and ancient writings of the Chinese, Egyptians, Eskimos, Hindus and others tell about the great opening in the north. The wise men of these brown races also teach that there is a race of men UNDER the earth crust. Also that some of their ancestors came from the interior of the earth.

This drawing shows how the needle of the compass works while explorers are passing into the interior of the earth. Notice how the compass would lead them out again, they not knowing the earth was hollow.

hollow, and is warmer in the interior than on the exterior. As the wind passes out in the winter, it warms the atmosphere. If the earth is solid, neither science nor reason can furnish any rational theory why it should be warmer as one passes north. Every known theory is against such a conclusion. As soon as you adopt the belief that the earth is hollow, perplexing questions will be easily solved, the mind will be satisfied, and the triumph of sensible reasoning will come as a delight never to be forgotten.

"This volume is not written to entertain those who read for amusement, but to establish and prove, as far as proof can be established and proved, certain mighty truths hitherto not comprehended. One key will unlock all these mysteries. The problems to be solved are the following:

"1. Why is the earth flattened at the poles?

"2. Why have the poles never been reached?

"3. Why is the sun invisible so long in winter near the farthest points north or south?

"4. What causes the Aurora Borealis?

"5. Where are the icebergs formed and how?

"6. What produces the many tidal waves in the Arctic?

"7. Why do meteors fall more frequently near the Poles and from where do they come?

"8. What causes the great ice pressure in the Arctic Ocean during still tide and calm weather?

"9. Why is there colored snow in the Arctic region?

"10. Why is it warmer near the Poles than 600 to 1,000 miles away from them?

"11. Why is ice in the Arctic Ocean frequently filled with rock, gravel, sand, etc.?

"12. Why does the compass refuse to work near the Poles?

"Should I be able to give reasonable answers to the above questions—answers that will satisfy any intelligent person—the public will admit, I believe, that I have fulfilled my task.

"I wish to acknowledge my indebtedness to the brave men who have spent their time, comfort and, in many cases, have given their lives, so that all may know the truth and geography of this wonderful planet. Through their reports I am able to prove my theory that the earth is not only hollow, but suitable in its interior to sustain human life with as little discomfort as on its exterior, and can be made accessible to mankind with one-fourth the outlay of money, time and life that it costs to build the subway in New York City. The number of people who can settle in this new world (if not already occupied) will be billions.

"I claim that the earth is not only hollow, but that all, or nearly all, of the explorers who spent much of their time past the rim of the polar opening have had a look into the interior of the earth. When Lieutenant Greely was beholding the mock sun at 120 degrees latitude, he was looking into our sister world in the earth's interior."

Reed answers the above questions as follows:

"1. Why is the earth flattened at the Poles? As the earth is hollow, it could not be round, is the answer. The opening to the interior would detract from its roundness in proportion to the size of the opening.

"2. Why have the Poles never been reached? Because no Poles exist in the sense usually understood.

"3. Why does the sun not appear for so long a time in winter near the supposed Poles? Because during the winter the sun strikes the earth obliquely near the Poles. As one passes over the rim of the polar opening and approaches the earth's interior, one sinks inward into the hollow interior. The sun's rays are in this way cut off, and do not appear again until they strike that part of the earth more directly and shine down into the opening. This explains why nights are so long in the far north.

"4. Assuming that the earth is hollow, the interior should be warmer. We will furnish evidence to prove that it is warmer. The ones who have explored the furthest north will be the best judges.

"5. Meteors are constantly falling near the supposed poles. Why? If the earth be solid, no one can answer this question. If the earth is hollow, it is easily answered. Some volcano is in eruption in the interior of the earth, and from it rocks are thrown into the air. Vast quantities of dust are constantly found in the Arctic Ocean. What causes this dust? The volcanic eruptions. The dust has been analyzed and found to consist of carbon and iron, which must come from some volcano in the polar opening.

"6. What produces the aurora borealis? It is a reflection of a fire within the interior of the earth. [According to Marshall B. Gardner, this fire is the central sun, whose rays project through the polar opening on the night sky, and the changing forms and streamers of the aurora borealis are due to passing clouds cutting off its rays.]

"7. Where are the icebergs formed? And how? The answer is as follows: In the interior of the earth, where it is warm, rivers flow to the surface through the polar opening. When they reach the outside, in the Arctic Circle, where it is very cold, the mouth of the rivers freezes forming icebergs. This continues for months, until, due to the warmer weather in summer and the warmth from the earth, the icebergs are thawed loose and are washed into the ocean. (The fact that icebergs are formed from fresh water, not salty ocean water, proves this theory.)

"8. What causes tidal waves in the Arctic? They are started by icebergs leaving the place where they are formed, and plunging into the ocean. This answer is given because nothing else can produce even a fraction of the commotion of a monster iceberg when it plunges into the sea.

"9. What causes colored snow in the Arctic region? There are two causes. The red, green and yellow snow are caused by a vegetable matter permeating the air with such density that when it falls with the snow it colors it. This vegetable matter is supposed to be the blossom or pollen of a plant. As it does not grow on earth, one can naturally believe that it grows in the interior and comes out through the polar opening. Black snow, often noticed, is caused by black dust, consisting of carbon and iron, and comes from a burning volcano. As no burning volcano is near the Arctic Ocean, it must be in the interior of the earth.

"10. Why is the ice filled with rock, gravel and sand?

These substances came from an exploding volcano near where the iceberg is formed.

"By treating the earth as hollow, we have the solution of all the great mysteries—such as tidal waves, ice pressures, colored snow, open Arctic Ocean, warmer north, icebergs, flattening of the earth at the Poles, and why the Poles have not been found, the supernatural giving way to the natural, as it always does with understanding, and relief comes to mind and body.

"The earth is hollow. The Poles so long sought are but phantoms. There are openings at the northern and southern extremities. In the interior are vast continents, oceans, mountains and rivers. Vegetable and animal life are evident in this new world, and it is probably peopled by races yet unknown to dwellers upon the earth's exterior."

In support of his theory of a hollow earth, Reed offers the following evidence:

LONG ABSENCE OF SUNLIGHT DURING LONG ARCTIC WINTERS. Reed summarizes the experience of Arctic explorers who very quickly passed from the region of sunshine into the region of long nights, or the opposite. In the far north the sun is absent for abnormally long perriods of time, which could not be the case if the earth was round and solid, or even just slightly flattened at the poles. The only explanation is that these explorers entered into the opening at the North Pole; and as they entered, the sun's rays were cut off from them, to reappear only when it was high enough in the sky to shine in.

ABNORMAL WORKING OF THE COMPASS IN THE FAR NORTH. This was observed by all explorers who reached very far north. This strange action of the compass is exactly what should be the case if the earth is hollow and if they entered into the polar opening. In his book Reed has a drawing of a cross-section of the polar opening with ships sailing both in and out. When the ship enters the polar opening, the needle of the compass assumes a vertical position, instead of horizontal, as it does on top of the earth's surface. This is due to entering the polar opening. This is exactly what explorers found to occur in the far north. They found that as they approached the pole, the needle of the compass becomes restless, and when one goes far enough north, assumes a vertical position, indicating that one has then entered the polar opening, as occurred with Nansen and others.

PASSING OVER THE RIM OF THE POLAR OPENING INTO THE EARTH'S INTERIOR. Reed says on this subject:

"Whenever the explorers pass into the interior, they meet such different conditions that they are puzzled to account for them. Therefore it is no wonder that they call it *a strange land*. Everyone who has spent considerable time in the Arctic or Antarctic Circles has met with conditions unexplainable according to the theory that the earth is round and solid—but which find an easy explanation according to the theory that it is hollow with openings at the poles. Greely's description of passing around the curve into the polar opening is exceedingly good and clear. He says:

" 'The deep interest with which we had hitherto pursued our journey was now greatly intensified. The eye of civilized man had never seen, or his feet trodden, the ground over which we were traveling. A strong, earnest

desire to press forward at our best speed seized us all. As
we neared each projecting spur of the lands ahead, our
eagerness to see what was beyond became so intense at
times as to be painful. Each point we reached brought
a new landscape in sight, and always in advance was a
point which cut off a portion of the horizon and caused
a certain disappointment.'

"If Greely and his companions were entering into the
interior of the earth, they would certainly find that the
earth has a greater curve near the poles than at any other
place; and as they passed over and around the farthest
point north, each projection reached would be followed
by another which always seemed to take in part of the
horizon. This is just what happened."

ROCKS IN ICEBERGS, COLORED SNOW, POLLEN AND DUST
IN THE FAR NORTH. On this subject Reed says: "When
it can be shown that conditions are such that no Arctic
icebergs (composed of fresh water) can be formed in the
far north on the earth's outer surface, they must be
formed in the interior. If the material that produces
colored snow is a vegetable matter (which the analysis
shows), and is supposed to be a blossom or the pollen of
a plant, when none such grows in the vicinity of the
Arctic Ocean, then it must grow in the interior of the
earth; for if it grows elsewhere on earth, then the snow
would be colored in other locations as well (as it is in the
vicinity of the polar opening), which does not seem to
be the case. The dust, so annoying in the Arctic Ocean,
is also produced by volcanic eruptions. Being light, it is
carried far away by the wind, and when it falls on ships,
it is disagreeable. When it falls on the snow it produces

black snow. When analyzed it is found to consist of carbon and iron, supposed to come from a burning volcano. Where is that volcano? No record or account of any near the North Pole is found; and if it be elsewhere, why does the dust fall in the Arctic Ocean?

"Various explorers report large rocks and boulders on and imbedded in the icebergs. These boulders are either cast there by the exploding volcano or they are scraped up as the bergs slide down the rivers in the interior of the earth. The dust in the Arctic is so heavy that it floats in great clouds. It colors the snow black; and it falls on ships in such abundance that it is a source of irritation. Nansen declares that it was one of his principal reasons for wanting to go home. If the earth is solid, there is no answer to this perplexing problem. But if the earth be hollow, the eruptions of volcanoes in the interior can easily account for the dust."

OPEN WATER AT THE FARTHEST POINT NORTH. "It is claimed by many that the Arctic Ocean is a frozen body of water. Although it always contains large bodies of drift-ice and icebergs, it is not frozen over. The student of Arctic travels will invariably find that explorers were turned back by open water, and many instances are cited where they came near being carried out to sea and lost. What I wish to present to the reader, however, is the proof that the Arctic Ocean is an open body of water, abounding with game of all kinds, and the farther one advances, the warmer it will be found. There are many cases of clouds of dust and smoke. Many fogs are reported in winter time. If the earth were solid, and the ocean extended to the Pole, or connected with land surrounding the Pole, there could be nothing to produce that fog. It is caused by the warm air coming from the interior of the earth. Kane (an Arctic explorer) writes: 'Some circumstances which he (McGary) reports seem to point to the existence of a north water all the year round; and the frequent water-skies, fogs, etc., that we have seen to the southwest during the winter, go to confirm the fact.'

"There are many pages of reports (in the writings of Arctic explorers) of this open sea to the far north. Greely speaks of open water the year round. If there be open water the year round at the farthest point north, can any good reason be assigned why all have failed to reach the Pole? The men who spent their time, comfort and, in several cases, their lives, were men more than anxious to succeed, yet, strangely, all failed. Was this because the weather got warmer and they found the game more plentiful? No, it was because there is no such place."

Nansen, who probably went farther north than any other explorer, remarks in his book that it was a strange feeling to be sailing in the dark night to unknown lands, over an open rolling sea, where no ship had ever been before, and remarks how mild the climate was for September. The farther north he went, the less and less ice he saw. He remarked, "There is always the same dark sky ahead, which means open sea. They little think at home in Norway that we are sailing straight to the Pole in clear water. I shouldn't have believed it myself if anyone should have predicted it two weeks ago, but it is true. Is this not a dream?"

Three weeks later he mentions that the water was still

open and not frozen. He remarks: "As far as the eye can see from the crow's nest with the small field glass, there is no end to the open water." Between September 6th and 21st, he found no ice as he traveled northward in a very high latitude.

Reed comments: "After all the foregoing evidence, is it possible that anyone can believe that the respective oceans (in the far north) are frozen bodies of water? If they do not believe that these oceans are frozen, why do the explorers fail to reach the Poles—if there be such places?

"WHY IT IS WARMER NEAR THE POLES. One of the principal proofs that the earth is hollow is that it is warmer near the Poles. If it can be shown by quoting those who made the farthest advance toward the supposed Poles, that it is warmer, that vegetation shows more life, that game is more plentiful than farther south, then we have a reasonable right to claim that the heat comes from the interior of the earth, as that seems to be the only place from which it could come.

"In 'Captain Hall's Last Trip,' we read: 'We find this a much warmer country than we expected, bare of snow and ice. We have found that the country abounds with life, and with seals, game, geese, ducks, musk-cattle, rabbits, wolves, foxes, bears, partridge, lemmings, etc.' (He is speaking of the far north.)

"Nansen draws special attention to the warmth and says, 'We must almost imagine ourselves at home.' This was at one of the farthest points north reached by anyone, and yet the weather was mild and pleasant.

"It will be observed that these extremely strong winds

from the interior of the earth not only raise the temperature considerably in the vicinity of the Arctic Ocean, but affect it very materially four hundred and fifty miles away. Nothing could raise the temperature in such a manner, except a storm coming from the interior of the earth.

"Greely states: 'Surely this presence of birds and flowers and beasts was a greeting on nature's part to our new home.' Does that sound as if he had expected to find these things there, or that their presence was an everyday occurrence? No. It was written in a tone of surprise. From what place had these birds and game come? South of them for miles, the earth was covered with perpetual snow—in many locations thousands of feet deep. They are found in that location in summer; and as it is warmer farther north, they would not be likely to go to a colder climate in winter. They seem to pass into the interior of the earth. The mutton-birds of Australia leave that continent in September, and no one has ever been able to find out where they go. My theory is that they pass into the interior of the earth via the South Pole."

Reed points out that many animals inhabiting the far north, as the musk-ox, go north in winter in order to reach a warmer climate. He remarks: "Since it becomes warmer as they go north, instinct tells them not to go south in winter. And if they do not go south, they must go into the interior of the earth."

Another animal that goes north in winter is the auk. Schwatka saw a flock of four million auks, which darken the sky, going north as winter approached. Nansen says of the extreme north that a land which teems with bears, auks and black guillemots "must be a Canaan, flowing with milk and honey."

Reed continues:

"WHAT PRODUCES COLORED SNOW IN THE ARCTIC? Why is the snow colored in the Arctic regions? The snow has been analyzed and the red, green and yellow have been found to contain vegetable matter, presumably a flower, or the pollen of a plant. From where did it come? A flower that produced pollen sufficient to permeate the air with such density that it colored the snow, which require a vast territory—millions of acres—to grow it. Where is that to be found? It must be near the North Pole, for, if it grew elsewhere, colored snow would be found at other locations, and not be confined to the Arctic regions. As no such flowering plant is known on the earth's surface, we must look elsewhere.

"The interior of the earth is the only spot that will furnish us with an answer to the question. As the colors fall at different seasons, we may presume that the flowers mature at these seasons. It is also easy to find out where the black snow, frequently mentioned by the explorers, comes from. It comes out of an exploding volcano—of the kind that covered Nansen's ship with dust. All unexplained questions could be easily answered if one would believe that the earth is hollow. It is impossible to answer them under any other theory.

"Kane, in his first volume, page 44, says: 'We passed the Crimson Cliffs at Sir John Ross in the forenoon of August 5th. The patches of red snow from which they derive their name could be seen clearly at the distance of ten miles from the coast. It had a fine deep rose hue.'

"Kane speaks of the red snow as if it had a regular season in which to appear—as he says, 'if the snowy surface were more diffused, as it is no doubt earlier in the season.' In another place he speaks of the red snow being two weeks later than usual. Now taking the fact into account that the material that colors the snow is a vegetable matter, supposed to be the blossom or pollen of a plant, and that no such plant grows on earth, where does it come from? It must grow in the interior of the earth."

WHERE AND HOW ARE ICEBERGS FORMED? Since icebergs are formed from fresh water, not salty ocean water, they could not be formed from the Arctic Ocean, but by some fresh body of water. However there is no fresh body of water in the polar region. Reed's theory is that icebergs are formed from rivers coming from the interior of the earth and flowing toward the surface through the polar opening. When they reach the cold exterior they freeze, while more water passes over the frozen part and freezes too, forming mountains of ice. With the coming of summer, these big masses of ice are thawed loose and break off, falling into the sea and producing the mysterious tidal waves observed in the far north. Reed says:

"It is simply out of the question for an iceberg to form in any location yet discovered. On the other hand, the interior of the earth—back from the mouth of rivers or canyons—being warmer, is just suited for the formation of icebergs. The mouth freezes first, and the river, continuing to flow to the ocean, overflows the mouth, and freezes for months, until spring. As the warm weather of summer advances, and, owing to the warmth of the earth, the bergs are thawed loose, and water from the rains in the interior rushes up, and they are shoved into the ocean, and tidal waves started.

"Note the difference. On the outside of the earth, the whole length of a stream is frozen, and the farther inland the harder the freezing, while in the interior of the earth (at the polar opening) only the mouth is frozen. In the interior of the earth, there is not only plenty of water to produce icebergs, but plenty to shove them into the ocean.

"For the last three hundred years a fairly steady stream of explorers have been trying to reach the Pole—Arctic and Antarctic—and no one has ever seen an iceberg leaving its original location and plunging into the ocean. Isn't it strange that no one thought of asking about their place of origin?"

In support of the theory that icebergs, made from fresh water, cannot be formed on the outside of the earth and must come from fresh water rivers in its interior, Reed quotes Bernacchi who, writing on his observations in the Antarctic, says: "There was less than two inches of rainfall in eleven and one-half months, and while it snowed quite frequently, it never fell to any great depth. Under such conditions, where would materials be found to produce an iceberg? Yet the greatest one on earth is there—one so large that it is called the Great Ice Barrier, rather than an iceberg—being over four hundred miles long and fifty miles wide. It is grounded in two thousand one hundred feet of water, and extends from eighty to two hundred feet above water." Reed comments:

"Now it would be impossible for this iceberg to form in a country having practically no rain or snow. As icebergs are made from frozen water, and there is no water to freeze, it evidently was formed at some place other than where it now is. The iceberg itself, being of fresh water, lies in an ocean of salt water.

"How do I know that the great ice barrier came from the interior of the earth? Or from the kind of river described? First, it could not come from the exterior of the earth, since icebergs are not formed there. That river must have been 2,500 feet deep, fifty miles across and from four to five hundred miles long, for these are the present dimensions of the iceberg. The river had to be straight or the iceberg could not pass out without breaking. It passed through a comparatively level country because the surface is still flat. Another proof that the interior of the earth is level near the Antarctic entrance is that many of the icebergs found in the Antarctic are long and slim. They are called 'ice tongues,' which indicates that they came out of rivers running nearly on a level. The icebergs found in the Arctic, on the other hand, are more chunky, indicating that they come from a more mountainous country, where the fall of streams is more abrupt, causing the icebergs to be shorter and thicker.

"When Bernacchi was voyaging in the Antarctic, he wrote: 'During the next two days we passed some thousands of icebergs, as many as ninety being counted

Marshall B Gardner

Author of The Theory of a Central Sun Within The Earth's
Interior

from the bridge at one time. There was very little variety of form among them, all being very large and bounded by perpendicular cliffs. There was a large quantity of fresh water at the surface, derived from the number of icebergs.'

"How does this account accord with your notions of how icebergs are formed in a country where Bernacchi reports less than two inches of rainfall in the whole year, and but small quantities of snow? Where is the water to come from that will produce such great quantities of icebergs averaging a thousand feet in thickness, and many of them several miles long? Those icebergs were on their way north—never to return—yet the ocean will always be filled with them, as others will come from the place where they came. Where is that place? There is no rain or melted snow to furnish the water to freeze into an iceberg. Icebergs can come from only one place—the INTERIOR of the earth.

TIDAL WAVES. Reed here repeats the description of Arctic tidal waves by various explorers. They lift the ice of the great ice fields to great heights and can be heard for miles in the distance before they reach the ship and for miles after they pass beyond the ship. Arctic explorers describe these tidal waves as follows: "Giant blocks pitched and rolled as though controlled by invisible hands, and the vast compressing bodies shrieked a shrill and horrible sound that curdled the blood. On came the frozen waves. Seams ran and rattled across them with a thundering boom, while we watched their terrible progress." Reed says: "These tidal waves are caused by some tremendous agency and I can think of

nothing more powerful than the plunging of an iceberg
into the ocean. The great frequency of these powerful
tidal waves seems to exclude the possibility of their being
caused by underwater volcanic eruptions."

VIEW OF THE WATER-SKY

The skies in the Arctic and Antarctic circles reflect the surface of the earth,
water and ice, accurately. No great enterprise is undertaken without first
consulting the water-sky.

Chapter IV

MARSHALL B. GARDNER'S BOOK, "A JOURNEY TO THE EARTH'S INTERIOR OR HAVE THE POLES REALLY BEEN DISCOVERED?"

Marshall B. Gardner spent twenty years in research, based on the reports of Arctic explorers, supplemented by astronomical evidence, before publishing, in 1920, his great book, "A Journey to the Earth's Interior or Have the Poles Really Been Discovered?" He did not seem to know about Reed's book and theory, so that both men developed their theories independently. Gardner's great contribution is the theory of a central sun, which is the source of the higher temperature in the region of the polar orifice and the aurora borealis, which Reed attributes to volcanic eruptions. A central sun as a source of heat and light makes possible the existence of plant and animal life in the earth's interior, also human life, which Reed believed to be a fact, but could not explain according to his theory, which did not include a central sun as a source of light, without which there could be no life.

Gardner also claims, and in his book presents astronomical evidence to prove, that not only the earth, but all planets of the solar system, have hollow interiors and central suns, which he traces to their original formation from a whirling nebula. As a result of centrifugal force, their rotation during their early formation when yet molten caused their heavier constituents to be thrown toward the outside, forming a solid crust on the outer surface of each planet and leaving the interior hollow, while a portion of the original fire remained in the center to form the central sun. Also the force of their rotation and movements through space caused openings to form at their polar extremities.

Why have Reed's and Gardner's books become so rare that it is practically impossible to obtain copies and they are not found in most libraries? Because they prove that there exists a large area not recorded on any map, which is not only equal to, but perhaps greater than the entire land area of the earth's surface—this uncharted land area being on the inside of the earth's crust. Naturally any government that learned about this vast territory would have ambitions to be the first to discover it and claim it, for which reason it would make every effort to keep this information secret, so that no other government might learn about it and claim this territory first. Since the United States Government was the first to learn about it as a result of the visit of Admiral Byrd, who flew for 1,700 miles into this "mysterious land beyond the Pole," which is not shown on any map, and saw mountains, forests, green vegetation, rivers, lakes and animals there, we can understand the reason for secrecy and why the books of two American writers Reed and Gardner, were suppressed and forgotten, in order to guard this secret.

EVIDENCE FROM ARCTIC EXPLORATION

Gardner's book is 450 pages in length. With fifty books, chiefly on Arctic exploration, in his bibliography, he was most thorough in his research. Gardner claimed that the earth is a hollow shell approximately 800 miles thick in its crust, with an opening at the polar end approximately 1,400 miles across. He says that the mammoth comes from the interior and is still living there, and the huge tropical animals found frozen in ice in the polar region were not prehistoric but were animals from the interior that came to the surface and were frozen in ice when they did. In support of his theory of a polar opening and a central sun in the hollow interior of the earth, Gardner points out that birds and animals migrate to the north in winter to find warmer weather. He also notes that when explorers go north of 80 degrees north latitude, they find the water to become warmer due to warm currents coming from the polar region, and the air becomes warmer due to warm winds from the north. These cause the open sea, in place of ice, in the extreme north. They also find red pollen on icebergs and glaciers, and find logs and other debris washed ashore by these warm currents from the north. Gardner summarizes the evidence in favor of his theory of a hollow earth with two polar openings and a central sun as follows:

"How do scientists explain the fact that when we go north it becomes colder up to a certain point and then begins to get warm? How do they explain the further fact that the source of this warmth is not any influence from the south but a series of currents of warm water and of warm winds from the north—supposed to be a land of solid ice? Where can these currents come from? How could they come from anything else but an open sea? And why should there be a warm open sea at the very place where scientists expect to find eternal ice? Where could this warm water possibly come from?

"Why also should explorers find the inhospitable ice cliffs of the far north covered in large areas with the red pollen of an unknown plant? And why should they find the seeds of tropical plants floating in these waters— when they are not found in more southern waters? How should logs and branches of trees, sometimes with fresh buds on them be found in these waters, all being borne down by the warm currents from the north?

"Why should the northern parts of Greenland be the world's greatest habitat of the mosquito, an insect which is only found in warm countries? How could it have gotten to Greenland if it came from the south? Where do all the foxes and hares go which are seen traveling north in Greenland? Where did the bears go? Was it possible that such large creatures as bears could find sustenance on plains of eternal ice?

"How do scientists explain the fact that practically every competent explorer from the earlier days down to Nansen has admitted that when he got to the Far North his theories of what he should find failed to work and his methods of finding his positions also failed to work? How do scientists explain these passages from Nansen which we have quoted, showing that he was absolutely lost in the Arctic region?

"How do scientists explain the migration of those birds which appear in England and other northern countries one part of the year, in the tropics in another part of the year, but disappear entirely in the winter? How do they explain the fact that neither Peary nor Cook was able to prove the claim of reaching the north pole? Even supposing both men to have acted in good faith is it not obvious that both were lost? How else explain the discrepancies in Peary's own narrative?

"Why, says the reader, did Peary not discover that immense orifice at the polar extremity of the earth if it was there?

"The reason is very simple and can best be explained by asking another question.

"Why did not man discover by looking around him, that he was living on the surface of what is, practically speaking, an immense sphere (to be exact, spheroid)? And why did man for centuries think that the earth was flat? Simply because the sphere was so large that he could not see the curvature but thought it was a flat surface, and that he should be able to move all over the surface of it appeared so natural that, when scientists first told him it was a sphere he began to wonder why he did not fall off, or at least, if he lived in the Northern Hemisphere, he wondered why the Australians did not fall off—for he had no conception of the law of gravity.

"Now, in the case of the polar explorers the same thing is true. They sail up to the outer edge of the immense polar opening, but that opening is so vast, considering that the crust of the earth over which it curves is eight hundred miles thick, that the downward curvature of its edge is not perceptible to them, and its diameter is so great—about 1,400 miles—that its other side is not visible to them. So, if an explorer went far enough he could sail right over that edge, down over the seas of the inner world and out through the Antarctic orifice, and all that would show him what he had done would be that as soon as he got inside he would see a smaller sun than he was accustomed to—only to him it might look larger owing to its closeness—and he would not be able to take any observations by the stars because there would be neither stars nor even a night in which to see them.

"But, says the reader, would not the force of gravity pull the explorer who got inside the orifice away from the surface into the central sun; for does not gravity pull everything to the center of the earth?

"The answer to this is, that in gravitational pull it is not the geometrical position that counts. Center, in the geometrical sense of the word, does not apply. It is the mass that attracts. And if the great mass of the earth is in its thick shell, it is the mass of that shell that will attract, and not a mere geometrical point which is not in the shell at all, but 2900 miles away from it, as is the approximate distance between the central sun and the inner surface of the earth. As a matter of fact it is the equal distribution of the force of gravity all through the shell that keeps the sun suspended in the spot which is equidistant from every part of the shell. When we are on the outside of the shell it is the mass of the shell that attracts us to its surface. When we go over to the inside of the shell that same force will still keep our feet solidly planted on the inner surface.

"We shall see all that when we explore the Arctic in earnest, as we shall easily be able to do with the aid of airships. And when once we have seen it we shall wonder why it was that for so long we were blind to evidence which, as is shown in this book, has been before men's eyes for practically a whole century and over."

Twenty-seven years after Gardner wrote this, Admiral Byrd did exactly what he hoped would be done. He flew by airplane into the north polar opening for 1700 miles and came to a land of trees, as Gardner believed would exist there, and also a warmer climate, as shown by the rivers, lakes, vegetation and animal life he observed there.

Gardner writes: "That the musk-ox is not the only animal to be found where we should hardly expect it, is evident from a note in Hayes' diary. When he was in latitude 78 degrees, 17 minutes, early in July, he said: 'I secured a yellow-winged butterfly, and—who would believe it—a mosquito. And also ten moths, three spiders, two bees and two flies.'"

Since these insects are not found further south, a land of ice and snow, the only explanation Gardner could offer for their origin is that they came from the interior of the earth through the polar opening.

Hayes' observations of insect life in the extreme north were confirmed by Greely, in his book "Three Years of Arctic Service," describing his observations in the Arctic, begun in 1881. In the preface to his book, Greely tells us that the wonders of the Arctic regions are so great that he was forced to modify his actual notes made at the time, and understated them rather than lay himself open to the suspicion of exaggerating. That the Arctic regions are so full of life and strange evidence of life farther north, that an explorer cannot describe it without being accused of exaggerating is surely a very strange thing if these regions only lead to a barren land of ever-lasting ice, as according to older geographical theories.

Greely reports birds of an unknown species, butter-flies, flies and temperatures of 47 to 50 degrees, also plenty of willow to make fires, and much fresh driftwood. He found two flowers different from any that he had ever seen.

In many pages of astronomical evidence, Gardner discusses the bright lights seen shining from the polar caps of Mars, Venus and Mercury, and concludes that these planets all have central suns and polar openings. He claims that the earth has the same and that the aurora borealis results from the projection of the rays of the central sun, passing through the polar opening, on the night sky. Gardner summarizes the evidence in favor of his theory as follows:

"As explorers go north of about 80 degrees north latitude, they find that the water, instead of becoming colder in the same ratio in which it had been getting colder as they left the temperate zone, gradually begins to get warm again, and they find that this warmth is brought down from the so-called frozen north in a warm current flowing from the polar regions. Furthermore they find that birds and animals migrate to the north to feed and breed, instead of to the south. In fact, when they get

into really high latitudes, explorers find a greater wealth
of animal and vegetable life than they do in the lower
latitudes of the Arctic and sub-Arctic regions. And as
they are sailing to these northern regions they find,
scattered on the icebergs and glaciers, the red pollen of
plants that grow—where? Only in the interior of the
earth. And they find logs and other debris of the land
washed down in these warm currents just spoken of.
And this is not all. In our chapter on the mammoth and
mastodon we shall adduce evidence to show that the
mammoth still lives in the interior—in fact we shall
exhibit case after case where the mammoth has floated
out from the interior incased in glaciers and icebergs and
has been frozen in crevasses in the interior near the polar
openings, and then carried over the lip by glacial movement into Siberia."

In addition to driftwood found in the extreme north,
whose origin, according to Gardner, could only be the
earth's interior, there are found trees with green buds in
the Arctic seas. Seeds of unknown tropical species have
also been found drifting down in the northern currents,
coming from the north, not the south. Among these was
the seed of the entada bean, a tropical seed, which was
found by a Swedish expedition near Trurengerg Bay.
Gardner comments: "This seed must have come from
the interior of the earth, for it is of a tree that only
grows under tropical conditions, and it would have been
disintegrated had it been drifting all over the world for
many months, as would be the case if it had come up
from the tropical regions of the exterior of the planet."

Sverdrup found so many hares around 81 degrees

north latitude that one inlet was called Hare Fiord.
There was also enough other game to keep the whole
exploring party well fed with meat.

Captain Beechey saw so many birds on the west coast
of Spitzbergen that the place reverberated with their
cries from dawn till dark. The little auk were so numerous
and so close together that sometimes a single shot killed
thirty of them. With sixteen birds to a cubic yard, there
were about four million of them. Rotgers were so numerous as to darken the sky, and their chorus could be
heard for four miles. There were also reindeer and
ducks. There were four varieties of seagulls over the
surrounding ocean, plus fish and amphibious animals,
from the huge whale to the minute clio on which it
feeds, swallowing perhaps a million with each mouthful.

Franklin saw large numbers of geese migrating to the
unknown north, at a high latitude, indicating land there.
He notes that no matter how far north the explorer goes,
he always finds the polar bear ahead of him. No matter
how far north these bears are met, they are always on
their way north.

At latitude 82, Kane found butterflies, bees and flies,
as well as wolves, foxes, bears, geese, ducks, water-fowls
and partridges. A strange fact all explorers observe is
that animals do not migrate south to escape the cold
Arctic winter, but instead go north.

Commander McClure explored Banks Land and
found immense quantities of trees thrown in layers by
glacious action, which evidently brought them from the
north. In one ravine he found a pile of trees closely
packed, to a height of forty feet. While some wood was

petrified, much of it was of recent origin. These trees were found far beyond the latitude where trees grow.

Nansen was puzzled by this driftwood which is continually found along the Greenland coast. He said that as far north as latitude 86 degrees he found such driftwood.

Gardner says that it is the unanimous testimony of explorers that "the further north you go, the more animal life there is, a complete proof that there is in the far north a great asylum of refuge where every creature can breed in peace and with plenty of food. And from that region must come also those evidences of vegetable life that explorers have repeatedly seen, the red pollen of plants that drifts out on favorable breezes and colors whole icebergs and glacier sides with a ruddy tinge, those seeds and buds and branches, and most impressive of all, those representatives of races of animals that yet live on in the interior, although they have disappeared from the outside of the earth. (Gardner here refers to mammoths found frozen in ice.)

"What a veritable paradise of animal and vegetable life that must be! And perhaps for some sort of human life, also, it is a land of perpetual ease and peace. The Eskimo people who are still living there will have been modified from the type that we see on the outer surface. Their life will be easier, as they will have no cold climates and food scarcities to contend with. Like the inhabitants of some of our tropical islands, they will reflect the ease of their lives in easy-going and lovable temperaments. They will be . . . eaters of many fruits and other vegetable products unknown to us. When we penetrate their

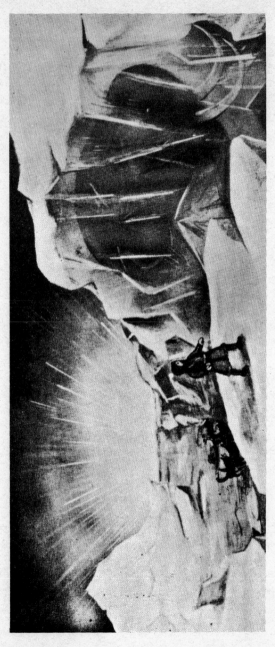

DISCOVERY OF THE MAMMOTH ENCASED IN ICE

Russian Fisherman of Tongoose, Siberia, in 1799, discovered a tremendous elephant, in a perfect state of preservation, as when it had died, enclosed in a huge block of ice, as clear as crystal. Though previously supposed to be a prehistoric animal which lived in the polar regions at a previous time when it had a tropical climate, according to the theory presented in this book, the elephant came from the earth's interior, which enjoys a tropical climate, and was frozen on reaching the exterior of the earth with the Arctic climate. (From Marshall B. Gardner's "A Journey to the Earth's Interior or Have the Poles Really Been Discovered?")

The central sun as it would appear to an explorer when he had reached the spot indicated by the letter "D" on the diagram, if the atmospheric conditions were favorable.

land we shall find growing almost to the inner edge of
the polar opening those trees of which we have seen so
many drifting trunks and branches. We shall find, nest-
ing perhaps in those trees, perhaps in the rocks around
the inner polar regions the knots and swans and wild
geese and ross-gulls that we have so often seen in the
preceding pages, flying to the north to escape the rigors
of climate which we in our ignorance have for so long
supposed to be worse in the north than elsewhere."

Speaking of Nansen, who reached further north than
any other explorer, Ottmar Kaub comments:

"Marshall B. Gardner was right when he wrote his
book in 1920. On August 3, 1894, Dr. Fridtjof Nansen
was the first man in history to reach the interior of the
earth. Dr. Nansen got lost and admitted it. He was
surprised at the warm weather there. When he found a
fox track, he knew he was lost.

"How could a fox track be there, he wondered. Had
he known that he had entered the opening that leads to
the hollow interior of the earth and that this was the
reason why, the further north he went, the warmer it
became, he would have found not only fox tracks but
later tropical birds and other animals, and finally the
human inhabitants of this 'land beyond the Pole,' into
which Admiral Byrd penetrated for 1,700 miles by
plane and which completely mystified him."

ORIGIN OF THE MAMMOTH

Gardner claims that the mammoth and elephant-like
creatures of tropical origin found frozen in the Arctic

ice, which is derived from fresh water (not salty water as one would suppose, since this is the only water found there) are really animals from the interior of the Earth that came to the surface and became frozen, and are not prehistoric animals as commonly supposed. Gardner's theory of the subterranean origin of the mammoth found confirmation in Admiral Byrd's observation of a living mammoth during his 1,700 mile flight into the land beyond the North Pole, within the polar opening.

Gardner claims that these strange animals not known on the Earth's surface were carried by rivers from the Earth's interior, freezing within the ice that was then formed. This theory seems very reasonable, in view of the ice being formed from fresh water not found in the Arctic Ocean. Since this ice, like icebergs, could not have been formed by ocean water, the only explanation is that it comes from *other* water—fresh water rivers flowing out through the polar opening from the earth's interior.

Since these animals are found inside of icebergs, which are composed of fresh water, this water, like the animals frozen in the ice it forms on reaching the surface and exposed to its lower temperature, must come from the earth's interior. Gardner speaks of herds of mammoths, elephants and other tropical animals which, when they venture out to the colder regions near the rim of the polar opening, together with glaciers which form there from water from the interior flowing outward and freezing, become frozen in the ice. Or they might fall into crevasses, perhaps concealed by snow, and the moment they fall in, they will be covered by snow and snow-water from above and hermetically sealed in the ice. This would account for the fresh condition in which these mammoths frozen in the ice are found after these glaciers have gradually worked their way over the rim of the polar opening and out into the Siberian wastes where these frozen animals are found in a perfectly fresh and edible condition.

Robert B. Cook tells of the remains not only of mammoths, but of hairy rhinoceros, reindeer, hippopotamus, lion and hyena, found in northern glacial deposits. He claims that these animals which were unable to endure cold weather were either summer visitors during the severity of the glacial period or permanent residents when the country had a milder climate. But Gardner maintains that these animals came from inside the earth for the following reason: "Since the reindeer, lion and hyena are present day forms of life and not as old as the mammoth (at least in the form in which we know them today and in which these remains show them to have been when they were alive), it is evident that these animals visited the spots where their remains were found not from southerly climates during early glacial epochs, but that they are remains of visitors from the land of the interior. Otherwise these present day forms would not be found alongside those of the mammoth which we have shown to be a present day inhabitant of the interior of the earth. Not knowing this, Mr. Cook has great difficulty in explaining the occurrence together of these forms which in his view are earlier and later forms of life. But when we shall see that they are really contemporaneous (and both came from the interior of the earth), the difficulty vanishes."

In the stomach of the mammoth was found undigested food consisting of young shoots of pine and fir and young fir cones. In others are found fern and tropical vegetation. How could an Arctic animal have tropical food in its stomach? One explanation is that the Arctic region once had a tropical climate, and that a shift of the earth on its axis suddenly brought on the Ice Age and changed the climate to a frigid one.

This theory has been offered to explain both the tropical vegetation in the stomach of frozen Arctic animals and the fact that many of these huge animals were of tropical species, related to elephants. Great deposits of elephant tusks were found in Siberia as evidence of the then northern habitat of tropical animals. But there is another theory to explain these facts: that these tropical animals came from the interior of the earth, which has a tropical climate, coming out through the North Polar opening. On reaching the cold exterior with its Arctic climate they froze, since they were unaccustomed to such cold climate. This is the theory held by Ray Palmer, who does not accept the idea that these animals died in prehistoric times as a result of a shifting of the earth on its axis. He says:

"True the death must have been sudden, but it was not because the Arctic was previously tropical and suddenly changed to a frigid climate. The sudden coming of the Ice Age was not the cause of death. The cause of death was Arctic in nature, and could have occurred any time, even recently. Since the Ice Age there were no mammoths in the known world, unless they exist in the mysterious land beyond the Pole, where one of them was actually seen alive by members of the Byrd expedition!

"We have taken the mammoth as a rather sensational modern evidence of Byrd's mysterious land, but there are many lesser proofs that an unknown originating point exists somewhere in the northern regions. We will merely list a few, suggestions that the reader, in examining the records of polar explorers for the past two centuries, will find it impossible to reconcile with the known areas of food mentioned early in this presentation of facts, those areas surrounding the polar area on your present-day maps."

ASTRONOMICAL EVIDENCE IN SUPPORT OF GARDNER'S THEORY OF A HOLLOW EARTH

Gardner devotes a considerable portion of his book to a discussion of astronomical evidence in support of his theory of a hollow earth with polar openings and a central sun by referring to the original formation of planets from nebulae and the polar lights observed from Mars, Venus and Mercury.

In reference to nebulae, Gardner points out that planetary nebulae show a shell structure, generally with a central star, as observed by H. D. Curtis of the Astronomical Society of the Pacific in an article in "Scientific American" on October 14, 1916. He reports:

"Fifty of these nebulae have been studied photo-

AURORA BOREALIS

There is nothing about this aurora, as described by Hall, that a great fire in the interior of the earth would not explain.

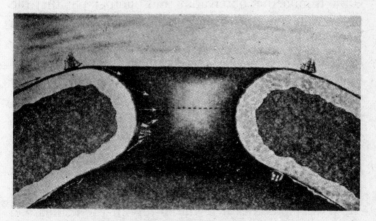

THE WORKING OF THE COMPASS

This illustration is presented to show how the magnetic needle works in passing into the interior of the earth, and how the compass would lead explorers out again, they not knowing the earth was hollow.

graphically with the Crosly reflector, using different lengths of exposure in order to bring out the structural details of the bright central portions as well as of the fainter, outlying parts. Most planetary nebulae show a *more or less regular ring or shell structure, generally with a central star.*"

On the basis of the above and other astronomical evidence, Gardner claims that the shape of the nebulae, as seen through the telescope, confirms his theory by showing that in the original formation of planets from nebulae, they acquire a hollow interior, polar openings and a central sun, as is indicated by the shape of the ring nebula shown on the accompanying photograph. Gardner writes:

"Why have scientists never really considered the problem of the shape of the planetary nebula? They know from actual observation and photographs that the planetary nebula takes the form of a hollow shell open at the poles and having a bright central nucleus or central sun at its center. Why have they never thought what that must imply? It is evidently one stage in the evolution of the nebula. Why have scientists never asked themselves what that conformation must logically lead to? Why do they ignore it altogether? Is it not because they cannot explain it without too great a disturbance of their own theories? But our theory shows how that stage in the evolution of a nebula is reached and how it is passed, we show what precedes it in the history of the nebula and what follows it. We show a continuous evolution passing through that stage to further stages in which those polar openings are fixed, the shell solidified, the nebula reduced to a planet. And it must be remembered that while the original nebula was incomparably greater than a planet in size, measuring even millions of miles across perhaps, at the same time that nebula is composed of gases so attenuated and so expanded by their immense heat that when they solidify they only make one planet."

Gardner points out that just as, in the formation of the solar system, some of the original fire remains at the center in the form of the sun, so, in the case of each individual planet, by the same process by which the solar system as a whole is formed, and by a continuation of the same general movement of rotation and the centrifugal throwing out of the heavier masses to the periphery (as shown by the fact that the most outermost planets, as Uranus and Neptune, are larger than those nearer the sun, as Mercury and Venus), in the case of each of the planets, in their formation, some of the original fire remains in the center of each, to form the central sun, while their heavier constituents are thrown to their surface to form the solid crust, leaving the interior hollow. Also, due to their rotation on their axis, centrifugal force causes the mass throughout to collect more at right angles to the axis of rotation, causing a bulge at the Equator, with a corresponding compensation at the poles in form of polar depressions which open to the hollow interior, rather than being perfectly round.

It is Gardner's theory, in support of which he presents astronomical evidence in his book, that all planets are hollow and have central suns, this being the basic pattern according to which solar systems are formed from the primordial nebulae from which they originate. Also our

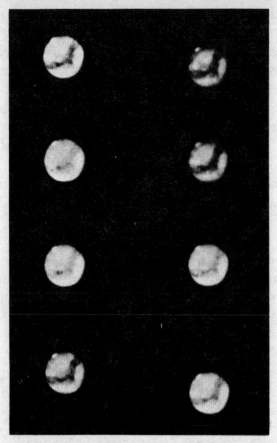

Views of Mars taken at the Yerkes Observatory, Sept. 28, 1902, showing the white circle or so-called snow-cap, projected beyond the planet's surface, which precludes all possibility of its being snow or ice. (*From Marshall B. Gardner's "A Journey to the Earth's Interior or Have the Poles Really Been Discovered?"*)

universe must have a central sun too, around which the stars circulate.

Gardner quotes the famous astronomer, Professor Lowell, that he has seen gleams of light from the polar cap of Mars. According to Gardner, this is due to the central sun of Mars passing through the polar opening. Similar bright lights have been observed coming from the polar region of Venus. During a transit of Mercury across the sun, the planet, while black on the side toward us, was observed to emit a bright light, comparable to the light of our sun, coming from its black disc.

Gardner concludes that these three planets are all hollow and have large polar openings misnamed polar

View of Mars, showing the circular white spot which is an entrance to this planet's interior, instead of the so-called polar ice cap, thus proving that Mars, the earth, and all other planetary bodies are hollow and contain a central sun. (Photographed by F. A. A. Talbott, Beighton, England.) For optical reasons all astronomical photographs are inverted.

A bird's-eye view of the opening to the interior of the earth.

caps of ice and snow, but in reality are white due to the large amount of fog and clouds in these regions, and that openings in the fog or clouds permit the central sun to shine through. Such bright lights have repeatedly been observed by astronomers who, not understanding the reason, could not offer any satisfactory explanation. Gardner notes that at times these polar caps disappear suddenly, due to a change of weather and that ice and snow could not melt so rapidly. Professor Newcomb says:

"There is no evidence that snow like ours ever formed around the poles of Mars. It does not seem possible that any considerable fall of such snow could take place, nor is there any necessity of supposing actual snow or ice to account for the white caps."

In support of his claim concerning the existence of lights seen at the pole of Mars Gardner quoted Professor Lowell who notes that on June 7, 1894, he was watching Mars and suddenly saw two points of light flash out from the middle of the polar cap. They were dazzling bright. The lights shone for a few minutes and then disappeared. Green, some years earlier, in 1846, also saw two spots of light at the pole of Mars.

Lowell tried to explain the lights he saw as reflections of sunlight by polar ice, but Gardner denies this, quoting Professor Pickering who saw a vast area of white form at the pole of Mars within twenty-four hours, visible as a white cap, and then gradually disappear. Also Lowell saw a band of dark blue, which he took to be water from the melting ice or snow cap. Gardner believes that the so-called Martian ice cap was really fog and clouds, which also could appear and disappear so rapidly. He writes:

"What Lowell really did see was a direct beam—two direct beams at the same moment—flashing from the central sun of Mars out through the aperture of the Martian pole. Does not the blue rim around that area to which Lowell referred indicate the optical appearance of the reflecting surface of the planet gradually curving over to the interior so that at a certain part of the curve it begins to cease reflecting the light? And the fact that it is not seen often simply shows that it is only visible when Mars is in a certain position with relation to the earth, when we are able to penetrate the mouth of the polar opening and catch the direct beam.

"Why have scientists never compared the facts of the

light cap of Mars with the light that plays over our own polar regions? Do they forget that the auroral display has been observed to take place without any reference to the changing of the magnetic needle? And if the aurora is shown to be independent of magnetic conditions, what else can it be due to than a source of light? Is not the reflection of the aurora light from the higher reaches of the atmosphere comparable to the projection of the light of the Martian caps into the higher reaches of the Martian atmosphere? And how do scientists explain the fact that the aurora is only seen distinctly in the very far north and only seen in a fragmentary way when we get further south?"

In support of his view that the polar caps of Mars are not formed of ice and snow but represent the light of its central sun shining through the polar opening, Gardner says:

"Why does the hot planet Venus have polar caps like those of Mars if the Martian caps are really composed either of ice, snow or frozen carbon dioxide? Also, why do the polar caps of Venus and Mercury not wax and wane as those of Mars are said to do? And why are the polar caps of Mars seen to throw a mass of light many miles above the surface of the planet when they are seen in a side view if they are really of ice? How could they be so luminous in the first place—more luminous than snow is when seen under similar circumstances? And how could Lowell see direct gleams of light from the caps if there were not beams from a direct light source?

"Furthermore, how do scientists account for the fact, noticed also by Professor Lowell, whose observations on

Mars all seem to support our theory, that when the planet is viewed through a telescope at night, that its polar light is yellow and not white, as the light from snow caps would be? The central sun is an incandescent mass, and just as the glowing of an incandescent electric light looks yellow when seen from a distance through darkness, so the direct light of the Martian sun would appear yellow—but if this light were reflected from a solid white surface it would certainly appear white. But it does not, and so it is up to the scientists to tell us just why it does not. But so far as we know they have not succeeded in doing this."

Mitchell saw two bright flashes of light at the polar cap of Mars which gradually came together. Gardner explains this as due to clouds which passed over the face of the interior sun, causing variations in the light emitted through the polar opening.

An English astronomer, W. E. Denning, writing in the scientific periodical, "Nature," concerning his observations in 1886, wrote:

"During the past few months the north polar cap of Mars has been very bright, sometimes offering a startling contrast to those regions of the surface more feebly reflective. These luminous regions of Mars require at least as much careful investigation as the darker parts. In many previous drawings and descriptions of Mars, sufficient weight has not been accorded to these white spots."

The English astronomer, J. Norman Lockyer, in 1892, wrote about Mars: "The snow zone was at times so bright that, like the crescent of the young moon, it appeared to project beyond the planet. This effect of irradiation was frequently visible. On one occasion the snow spot was observed to shine like a nebulous star when the planet itself was obscured by clouds, a phenomenon noticed by Beer and Madler, and recorded in their work, 'Fragments Sur les Corps Celestes.' The brightness seemed to vary considerably, and at times, especially when the snow zone was near its minimum, it was by no means the prominent object it generally is upon the planet's disc."

Gardner comments on the above observations:

"No one who reads the above in the light of our theory can fail to see how it fits in. Only direct beams of light from a *central sun* could give that luminous effect above the surface of the planet and varying as the atmosphere in the interior or above it was clouded or clear. Had it been a mere ice cap, there would not have been this luminosity when the planet was covered with clouds, as Lockyer says it was. Furthermore, that luminosity is precisely what our aurora borealis would look like if our planet was viewed from a great distance. And the light is the same in both cases. By turning to the planet Venus we shall demonstrate absolutely that the polar circles are not snow, or ice, or even hoar-frost caps, but simply apertures leading to the inner and illumined surface of the planet."

On Venus the extensive water vapor tends to equalize the temperature, so that its polar caps are not composed of ice and snow, as supposed in the case of Mars, but which Gardner doubts. Speaking of the polar caps of Venus, Macpherson, in his "Romance of Modern Astronomy," says:

"Polar caps have been observed, supposed by some to be similar to those on our own planet and Mars. Some astronomers, however, do not regard them as snow."

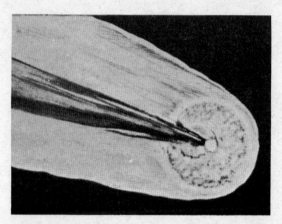

A Photographic Reproduction of a Drawing Showing the Head of Donati's Comet as Seen in 1853

Nothing could more strikingly support our theory than the above illustration. It is taken from a drawing of Donati's Comet, as seen from Cambridge Observatory on October 1st, 1853. The central nucleus is very plainly seen, surrounded by a sphere of glowing gases, and enclosed by an outer envelope. The comet is passing through an area of conflicting forces, and this, and perhaps the excessive heat of the body has caused the great split which extends through the envelope to the central sun itself. A comet is simply a planet which is disintegrating, and this photograph shows us the disintegration taking place, and just far enough advanced so that we can see the inner structure of the planet. And that structure is precisely what our theory says is the actual structure of all planets, our earth included. As the reader continues in this book let him bear this picture in mind, and he will be more and more struck by the happy manner in which the theory is verified by the structure here shown. And let him remember that this picture was not made up to support our theory, for it was made many years before our theory was promulgated.

The French astronomer Trouvelet, in 1878, observed at the pole of Venus a confused mass of luminous points, which Gardner attributes to light from the central sun struggling through the clouds. Since the polar cap is not made of ice, these lights cannot be a reflection of the sun. He believes this is the same case with Mars.

Similar lights are seen coming from Mercury. Richard Proctor, one of the best known astronomers of the nineteenth century, wrote: "One phenomenon of Mercury, if real, might fairly be regarded as indicating Vulcanian energies compared with which those of our own earth would be as the puny forces of a child compared with the energies of a giant. It has been supposed that a certain bright spot seen in the black disc of Mercury when the planet is in transit indicates some source of illumination either of the surface of the planet or in its atmosphere. In its atmosphere it could hardly be; nor could any auroral streamers on Mercury be supposed to possess the necessary intensity of lustre. If the surface of Mercury were glowing with the light thus supposed to have been seen, then it can readily be shown that over hundreds of thousands of square miles of that surface must glow with an intensity of lustre compared with which the brightness of the lime light would be as darkness. In fact, the lime light is absolute darkness compared with the intrinsic lustre of the sun's surface; and the bright spot supposed to belong to Mercury has been seen when the strongest darkening-glasses have been employed. But there can be no doubt that the bright spot is an optical phenomenon only."

A SECTION OF THE GREAT ICE BARRIER

A monster iceberg in the Antarctic Ocean, four hundred miles long, fifty miles wide, grounded in twenty-one hundred feet of water, and extending from one hundred to two hundred feet above the ocean; frozen from fresh water, not attached to land. How did it get there?

Commenting on Proctor's statement, Gardner writes: "Again we agree with the observation but not with the inference. Here is a spot of light on Mercury, plainly seen through a telescope, so bright that the observer compares it to the incandescence of a sun. It is a much brighter light than any reflection could possibly give. To Proctor such an appearance must have been shocking to the extreme. He was not expecting it and was utterly unprepared to see such a phenomenon. So he is utterly unable to explain it. So Proctor calls this light 'an optical phenomenon only.' But we cannot believe that Proctor's eyes have played him a trick. He was a trained astronomical observer. So what he saw must have had some explanation or cause behind it.

"It is obvious to us that what he saw was the central sun of Mercury beaming directly through the polar aperture, and as Mercury is a small planet, the interior sun would be rather near the aperture, and there would be no aqueous atmosphere with clouds to darken its beams, with the result that this sun would shine with extraordinary brightness. It may be noticed that its beams put Proctor in mind of the beams from the sun that shines upon all the planets.

"What more could be wanted than this to show that Mercury, as well as the other planets, has a central sun, and that such a sun is to be met with universally? Is it not significant that beginning with observations on Mars, we are able to go on to Venus and Mercury, apply the same tests and get the same results? The tests are direct observation or photographic observation. The results are the invariable appearance of a central sun."

In addition to the above astronomical evidence in favor of his theory, Gardner refers to the structure of the heads of comets, showing a hollow center, outer crust and central sun. In his book he presents a drawing of Donati's comet, detected from a Florence observatory in 1858. As can be seen it had a central nucleus or sun, which "shone with a brilliance equal to that of the Polar Star" and was 630 miles in diameter. Gardner believes that a comet is a planet which came into the orbit of some other larger body, like our sun, which tore it from its own orbit, and possibly collided with another planet and the resulting heat transformed most of it into a gaseous tail that trails after it. Gardner claims that the fiery nucleus of the comet was once the central sun of the planet from which it was formed after it broke into fragments.

ORIGIN OF THE AURORA BOREALIS

Just as there are polar lights from Mars, Venus and Mercury, coming from their central suns shining through their polar openings, so Gardner claims, the same occurs in the case of our own planet, the polar lights which it gives off being the aurora borealis, which is not due to magnetism but to the earth's central sun.

Gardner presents the following theory of the origin of the Aurora Borealis:

"Why have scientists never compared the facts of the light cap of Mars with the light that plays over our own polar regions? Do they forget that the auroral display has been observed to take place without any reference to the changing of the magnetic needle? And if the aurora is shown to be independent of magnetic conditions, what else can it be due to than a source of light? Is not the reflection of the aurora light from the higher reaches of the atmosphere comparable to the projection of the light of the Martian caps into the higher reaches of the Martian atmosphere? And how do scientists explain the fact that the aurora is only distinctly seen in the very far north and only seen in a fragmentary way when we get further south?"

Gardner concludes that the aurora borealis is due to the central sun shining through the polar orifice on the night sky; and the variations in the streamers of light are due to passing clouds in the interior, which, in their movements, cut off the light of the central sun and cause the reflection on the sky to keep changing. That the aurora is not due to magnetism or electrical discharges is proven by many observations of Arctic explorers showing there is no disturbance of the compass or crackling sounds that accompany electrical discharges, when the aurora is most intense.

Gardner says: "There are some other considerations which show that the aurora is really due to the interior sun. Dr. Kane, in his account of his explorations, tells us that the aurora is brightest when it is white. That shows that when the reflection of the sun is so clear that the total white light is reflected, we get a much brighter effect than when the light is cut up into prismatic colors. In the latter case the atmosphere is damp and dense (in the interior of the earth) —that being the cause of the rainbow effect—and through such an atmosphere one cannot see so much. Hence the display is not so bright as it is when the atmosphere is clear and the light not broken up.

"Again, if the aurora is the reflection of the central sun, we should expect to see it fully only near the polar orifice, and see only faint glimpses of its outer edges as we went further south. And that is precisely what is the actual fact of the matter. Says Dr. Nicholas Senn in his book, 'In the Heart of the Arctics':

" 'The aurora, which only occasionally is seen in our latitudes, is but the shadow of what it is to be seen in the polar region.'

"The aurora is not a magnetic or electrical disturbance but simply a dazzling reflection from the rays of the central sun. For if it warms continents and waters in the interior of the earth, if, as we have seen, birds have their feeding and breeding grounds there, if an occasional log or seed or pollen-like dust is seen in the Arctic that came from some such unknown place as we have described, it ought to be possible to obtain enough evidence of such life."

Chapter V

WAS THE NORTH POLE
REALLY DISCOVERED?

Returning from the Arctic in September, 1909, Dr. Frederick A. Cook announced that he had reached the North Pole on April 21, 1908. His announcement was followed a few days later by one from Rear Admiral Robert E. Peary, who said he had reached the North Pole on April 6, 1909. Each man hurled accusations against the other, claiming that he had discovered the North Pole and that the other did not. Cook accused Peary, saying that he had appropriated some of his reports on his return from the Pole. But Cook failed to produce any written record that he had made of his trip, and this made his reports seem suspicious.

Though Cook claimed to be the first to reach the North Pole, Peary is generally given credit to have been the first to discover it. Cook's claim was discredited because the sun's altitude was only a few degrees above the horizon and was so low at the time that observations of it as proof of his position were worthless. Peary reached, or claimed he reached, the North Pole in April, fifteen days earlier in the season, and hence under more adverse solar conditions. His calculations are therefore more open to suspicion than Cook's.

Also Cook had no witnesses that he found the North Pole, other than Eskimos. The same is true of Peary, who lacked witnesses through choice, having ordered the men on the expedition to remain behind, while he went on alone with one Eskimo companion to the Pole. While Cook was doubted when he claimed to make 15 miles a day, Peary claimed to have made over 20 miles. The argument whether Cook or Peary, or neither, discovered the North Pole is still not perfectly settled.

There is one factor in Peary's dash to the Pole that casts suspicion on his claim to have reached it. This was the remarkable speed at which he claimed to travel, or would have had to travel to reach the North Pole and return during the time he did. When he neared the 88th parallel north latitude, he decided to attempt a final dash to the Pole in five days. He made 25 miles the first day; 20 miles on the second day; 20 miles on the third day; 25 miles on the fourth; and 40 miles on the fifth. His five-day average was 26 miles a day. Can a man walk that fast under the incredibly difficult conditions of the North Pole area, supposedly an ice-terrain described by the men in the atomic submarine "Skate" as fantastically jumbled and jagged? And yet, further south, with presumably better conditions of travel, he was able to average only 20 miles a day.

From these facts we must conclude that *neither Cook nor Peary reached the true North Pole*, since, according to the theories presented in this book, it does not exist. What Cook and Peary reached was probably the magnetic rim of the polar opening or depression, where the compass points straight down, but not the Pole itself, which lies in the center of this opening. Peary may have traveled for the distance he calculated as correct to reach the North Pole, but what he really did was to travel this same distance either around or into the depression or opening which exists in this part of the world, into which Admiral Byrd entered; and the further he would travel the deeper he would go into this opening, without ever reaching the true Pole.

Scientific societies that considered Cook's and Peary's claims to reach the North Pole concluded that in neither case could it be said authoritatively that the explorer had reached the Pole.

Cook's claim to have reached the Pole was based on his promise to prove it by field notes and mathematical observations. But he was never able to present any notes. He claimed that Peary caused some of this data to be buried. But in time the faith in Cook turned into skepticism, which was started by Peary's denial of Cook's claim. Peary's denial was supported by Cook's failure to present proper scientific data. Rear Admiral Melville of the United States Navy, an old time Arctic explorer, said in an interview at the time:

"It was the crazy dispatches purporting to have come from Dr. Cook about the conditions he found there, and other things, that caused a doubt in my mind about Cook's having found the Pole."

According to Dr. Tittman, Cook and Peary could not have traveled on foot over solid ice to reach the North Pole, because practically all scientists agree that this is not the fact. Some think there is open sea there and others fertile land. All explorers who have gone far enough north found open sea. As for fertile land there, this would only be possible according to our own theory of a polar opening and central sun, since, according to the theory of a solid earth, it should get colder and colder the further north one goes. But Arctic explorers found the opposite to be true. They found it warmer near the Pole than further south. But even if the cold at the Pole was not enough to freeze the sea, how could it be warm enough to permit fertile land unless our theory is correct? Since all polar explorers agree that there is open sea in this region (the polar orifice), but ice further south, it is clear that Cook did not go as far north as he thought he went.

When the Swedish Academy of Sciences and University of Copenhagen examined Cook's claims, they decided that he had not proved that he reached the Pole. Peary gave the following report to the Associated Press:

"Cook was not at the North Pole on April 21, 1908, nor at any other time. Cook's story should not be taken too seriously. The two Eskimos who accompanied him say he went no distance north, and not out of sight of land. Other members of the tribe corroborate this story. He has simply handed the public a gold brick."

But when Peary returned to civilization his own story sounded as dubious as Cook's. He had taken even fewer observations of his alleged position than Cook had done. The fact that he left his white companions behind and had no witnesses cast doubt on his claims. When Cook was doubted when he said he made fifteen miles a day in sledge traveling, Peary claimed he made over twenty, and even forty. Since it is impossible to make forty miles a day on a dog sledge, which is admitted to be slower travel than on foot, this claim seems impossible. When questioned whether he traveled faster on the dog sledge than on foot, Peary admitted:

"In Arctic expeditions a man is lucky if he is able to walk without pushing the sledge. Usually he must grip the rear and push it ahead. It is like guiding a breaking plow drawn by oxen. You must also expect at any moment that the sledge may strike some pressure ridge that will wrench you off your feet."

According to Peary's statement it seems impossible that he could travel at speeds of twenty to forty miles a day over Arctic ice and keep it up for eight days after doing equally arduous work for months.

For this reason, after examining Cook's and Peary's data, one investigator concluded:

"The question whether Cook or Peary discovered the North Pole may never be solved. It seems to be one of history's puzzles, and to remain a matter of one man's word against another."

When Peary submitted his proofs for investigation, the Congressional Committee that examined them acknowledged in Congress that Peary had not, any more than Cook, proved his claim of reaching the Pole. Peary claimed he traveled a distance of 270 miles from eighty-seven degrees, forty-seven minutes North to the Pole and back to the same latitude in seven days and a few hours. This speed seems impossible in the polar region.

Cook admitted he did not reach the Pole in his book he wrote after he returned from his expedition, in which he wrote:

"Did I actually reach the North Pole? . . . If I was

mistaken in approximately placing my feet upon the pin-point (North Pole) about which this controversy has raged, I maintain it was the inevitable mistake any man must make. To touch that spot would be an accident."

This created an international scandal. After foreign kings and universities had congratulated and showered honors on Cook, later it was discovered they had been duped. Now, after one American explorer (Cook) was found to have made a false claim, it would reflect badly on the reputation of the United States if another (Peary) was found, after examination, to also make a false claim. This would lead to ridicule in the foreign press. To prevent this, the Congress of the United States appointed a committee of the National Geographical Society, which gave a favorable verdict on Peary's discovery after a cursory examination of his field notes, and it was hoped this would settle the matter, so that the world may consider an American explorer, Peary, to have discovered the North Pole. It was hoped this would settle the matter, and prevent one false claim about the discovery of the North Pole by an American from following the other.

However, a year after the committee of the National Geographical Society made a favorable verdict on Peary's claim, a new Congressional investigation was made and its verdict was that Peary did not prove his claims because his statements were not backed by a single white witness. The committee made the verdict of "not proven."

But Peary never replied to the charges made against him, and wished to end his career by retiring with the rank of Rear Admiral, which carried a pension with it of $6,000 a year. Friends of Peary brought into Congress a bill to retire him. One would think that before he retired an inquiry would be made whether or not he reached the Pole, but no inquiry was made. While the United States government refused to officially endorse Peary's discovery, it could not afford to lower its prestige before the world by announcing that he did not discover the North Pole.

"I am satisfied that Peary did not discover the Pole for two reasons:

"1. In spite of all the talk there has been about scientific data brought back by him and furnished as evidence, the fact is that his claim to the discovery in question is backed by his unsupported word, and by nothing else.

"2. All of the other claims to discoveries in the Arctic region by Peary have been proven false. Why, then, should we accept as true his unsupported statement that he arrived at the Pole?"

At a Congressional Hearing, Mr. Tittmann, superintendent of the U.S. Coast Survey, was asked: "What evidence is there that this party consisting of Peary and others, reached the Pole?"

Mr. Tittmann replied: "I have no evidence of that except the soundings recorded under Peary's signature. Peary brought back nothing—no witnesses, no worthwhile scientific proof, nothing but his unsupported word to back up his claim to have discovered the Pole. But, inasmuch as his reputation for veracity has been completely shattered by the fact that every other claim of discovery made by him has proven false, there is nothing that the world can accept as demonstrating that at any time he has been anywhere near the Pole."

Due to the irregular action of the compass in the polar region and the fact that the sun was barely above the horizon when both explorers were there, making it difficult to make measurements, in a region where it is easy for an explorer to get lost due to difficulty in ascertaining his position, it is probable that neither Cook nor Peary really found the North Pole, even if they thought they did. This is confirmed by the fact that every previous Arctic explorer found warmer conditions and open sea very far north, while Cook and Peary claimed they traveled over ice. This would indicate that they were in points further south and if they had gone further north they would have reached open sea. Concerning this, Marshall B. Gardner, in his book, "A Journey to the Earth's Interior or Were the Poles Really Discovered," writes:

"Had they (Cook and Peary) gone further they would have found open sea and increasing temperature. Had they then possessed boats they could have launched on that sea and the way to the goal and to the truth would have been clear. They would have seen the earth's central sun shining even in the winter, shining all of the twenty-four hours and all of the year, and they would have discovered new continents and oceans, a new world of land and water and of forms of life some of which have vanished from the outside of the globe.

"But it was not to be. The discovery of that new land was left to those who, following the theory outlined in this book, and using such safe means of Arctic traveling as the airplane and dirigible, will fly over the eternal barrier of ice to the warmer sea beyond and over that until they come into the realm of perpetual sunlight."

Gardner's claim was confirmed by the two expeditions of Admiral Byrd, which traveled by airplane through the openings at the North and South Poles and came to this warmer land, where they saw a new strange form of animal life, as well as trees, green vegetation, mountains and lakes, though the expeditions did not penetrate the polar openings far enough to reach the tropical land of perpetual sunlight in the earth's interior, about which Gardner speaks. But such a land and such a sun must exist if Admiral Byrd's observations of a warmer territory beyond the Poles are correct.

ROCK ON ICE

The author claims that the rocks shown in the above illustration were thrown into the air by a volcanic explosion, and dropped upon the berg while it was forming.

Chapter VI

THE ORIGIN OF THE ESKIMOS

William F. Warren, in his book, "Paradise Found, or the Cradle of the Human Race," presents the view that the human race originated on a tropical continent in the Arctic, the famed Hyperborea of the ancient Greeks, a land of sunshine and fruits, whose inhabitants, a race of gods, lived for over a thousand years without growing old.

The ancient writings of the Chinese, Egyptians, Hindus and other races, and the legends of the Eskimos, speak of a great opening in the north and a race that lives under the earth's crust, and that their ancestors came from this paradisiacal land in the Earth's interior. (May not Santa Claus represent a race memory of a benefactor of humanity who came from this subterranean race, who came to the surface through the north polar opening—perhaps on a flying saucer, symbolized by his flying sled and reindeer?)

Most writers on the subject claim that the interior of the earth is inhabited by a race of small brown-skinned people and also say that the Eskimos, whose racial origin differs from that of all other races on the earth's surface, came from this subterranean race. One explorer declared that those known as the Arctic Highlanders came from the interior of the earth. When the Eskimos were asked where their forefathers came from, they pointed to the north. Some Eskimo legends tell of a paradisiacal land of great beauty to the north. Eskimo legends also tell of a beautiful land of perpetual light, where there is neither darkness at any time nor a too bright sun. This wonderful land has a mild climate where large lakes never freeze, where tropical animals roam in herds, and where birds of many colors cloud the sky, a land of perpetual youth, where people live for thousands of years in peace and happiness. There is a story of a British king named Herla, whom the Skraelings (Eskimos) took to a land of paradise beneath the earth. The Irish have a legend about a lovely land beyond the north, where are continuous light and summer weather. Scandinavian legends tell of a wonderful land far to the north, called "Ultima Thule." Palmer comments: "Is Admiral Byrd's 'land of mystery,' 'center of the great unknown' the same as the 'Ultima Thule' of Scandinavian legend?"

Speaking of the origin of the Eskimo, Gardner says: "That the Eskimo came from the interior of the earth, that is to say, from a location which they could not easily explain to the Norwegians who might have asked them where they originally came from, is shown by the fact that the early Norwegians regarded them as a supernatural people, a species of fairy. When we remember that in the efforts of these Eskimos to tell where they came from they would point to the north and describe a land of perpetual sunshine, it is easy to see that the Norwegians who associated the polar regions with the end of the world, certainly not with a new world, would wonder at the strange origin thus indicated. They would naturally assume that these were supernatural beings who came from some region under the earth—as that was always considered to be the abode of fairies, gnomes and similar creatures."

And according to Nansen this is precisely what happened. He says:

"I have already stated that the Norse name 'Skraeling' for Eskimo must have originally been used as a designation of fairies or mythical creatures. Furthermore there is much that would imply that when the Icelanders first met with the Eskimo in Greenland they looked upon them as fairies. They, therefore, called them 'trolls,' an ancient common name for various sorts of supernatural beings. This view persisted more or less in later times."

Nansen goes on to tell us that when these Skraelings, or Eskimos, were mentioned in Latin writings, the word was translated as "Pygmaei," meaning "short, undergrown people of supernatural aspect." In the middle ages they were supposed to inhabit Thule, which refers to the ultimate land beyond the north. This belief in Thule, a land beyond the Pole, inhabited by a strange people, was very widespread. Nansen tells us that from St. Augustine the knowledge of these pigmies reached Isidore, and from him it passed over all of medieval Europe—in the sense of a fabulous people from the uttermost parts of the north, a fairy people.

A Welshman, Walter Mapes, in the latter part of the twelfth century, in his collection of anecdotes, tells of a prehistoric king of Briton called Herla, who met with the Skraelings or Eskimos, who took him beneath the earth. Many early legends tell of people going under the earth into a strange realm, staying there for a long period of time and later returning. The ancient Irish had a legend of a land beyond the sea where the sun always shone and it was always summer weather. They even thought that some of their heroes had gone there and returned—after which they were never satisfied with their own country.

A thirteenth century Norwegian writer is quoted by Nansen, according to whom the Eskimos were believed at this time to be a supernatural people, small in stature, and hence different in their origin than the other inhabitants of the earth. Gardner writes:

"Nansen says that Eskimo settlements increase not only by the tribe growing in numbers, but by 'fresh immigration from the north,' which clearly points to further additions from the interior of the earth.

"That they originally came from a land of constant sunshine, from a country much past the northern ice barrier is the tradition of the Eskimos themselves, and it is a tradition which must be given full weight, for it could not have arisen among them in the first place without cause. On this point Dr. Senn says: 'When questioned as to the land of their origin, they invariably point north without having the faintest perception what this means.'

"Naturally the Eskimos do not know that the earth is hollow and that ages ago they lived in its interior, but they have clung to that one simple fact—they came from the north. Dr. Senn denies that they have any characteristics in common with the North American Indian and thinks that they are the remnant of 'the oldest inhabitants of the western hemisphere.' In this attributing

of great antiquity to them he may be right—at least he
there agrees with Nansen. But the interior of the earth
and not the western hemisphere is evidently the place of
their original abode.

"As for the land of perpetual sunshine, the Eskimo,
of course, does not remember that as something he himself has seen, for it is very questionable if any of the
Eskimos of the present generation have ever penetrated
to the interior. But it is a well-known fact that every race
has its idea of a 'golden age' or paradise which is generally
composed of the elements being handed down in its
stories and myths as being characteristic of its earliest
home. Thus the Eskimo legends handed down generation after generation, tales of the interior land with its
ever shining sun—and what could be more natural than
when the Eskimo came to build in fancy a paradise for
himself and his loved ones after they should die, that
he should reconstruct this first home of which he had
heard only dim legends? That at any rate, is just what
he had done. Dr. Senn, discussing their religion says:

" 'They believe in a future world. The soul descends
beneath the earth into various abodes—the first of which
is somewhat in the nature of a purgatory. But the good
spirits passing through it find that the other mansions
improve till at a great depth they reach that of perfect
bliss, where the sun never sets, and where by the side of
great lakes that never freeze, the deer roam in large
herds and the seal and the walrus always abound in the
waters.'

"That paradise might serve as almost a literal description of the land in the interior of the earth, and the way

in which the Eskimo indicates a preliminary purgatory before it can be reached may be the reflection of a memory handed down in the tribe of the great hardships and difficulties of the ice barrier between that wonderful home and the present situation of the Eskimo on the southern side of that great natural obstacle.

"It is also interesting to note that when the Eskimo first saw Peary's effort to get further north than the great ice-cap of Greenland—beyond which they themselves had no ambition to explore—they immediately thought that the reason for his trying to get further north was to get into communication with other tribes there. That idea would hardly have occurred to them if it were not for the fact that they had traditional or other evidence of people in the supposedly unpopulated north.

"With such a weight of evidence all pointing one way it is very hard to resist the conclusion that in the Eskimo we find a type, changed now and mixed with other types, but still something of a type of human being that has inhabited or very likely still inhabits the interior of the earth. We can certainly find no origin for them that explains their present situation. And their legends admit of no other explanation either. For those legends certainly point to the same sort of land as every chapter of this book has pointed to—a land of perpetual sunlight and a mild climate, a land corresponding to the 'Ultima Thule' of ancient legend and that may, sooner than the skeptic expects, be opened up once more to those who go properly equipped to seek it."

Gardner says that both the Eskimo and Mongolian races came from the interior of the earth, since they resemble each other in many ways, including the unusual formation of their eyes, so different from that of other races. Gardner writes:

A SWARM OF AUKS

These birds are found so plentiful in the Arctic regions that when they fly overhead they darken the skies, their little voices being often heard from a distance.

PTARMIGAN

Birds found in great numbers in the Arctic Circle.

EIDER-DUCKS

Found in great numbers in the Arctic Circle.

"It is quite possible that the Eskimos are not descended from any tribes driven out of China as that might imply, but that the Chinese as well as the Eskimos originally came from the interior of the earth."

Chapter VII

THE SUBTERRANEAN ORIGIN
OF THE FLYING SAUCERS

Evidence That They Come From the Hollow
Interior of the Earth

The conception of a hollow earth presented in this book offers the most reasonable theory of the origin of the flying saucers and is far more logical than the belief in their interplanetary origin. For this reason, leading flying saucer experts, as Ray Palmer, editor of "Flying Saucers" magazine, and Gray Barker, a well known writer on flying saucers, have accepted the theory of their subterranean origin as against the idea that they come from other planets.

The theory that flying saucers came from the Earth's interior and not from other planets originated in Brazil and only later was it taken up by American flying saucer experts.

In 1957, while browsing in a Sao Paulo, Brazil, bookstore, the author came across a book that struck his attention, entitled, "From the Subterranean World to the Sky: Flying Saucers," by O. C. Huguenin. It was the book's thesis that flying saucers were not space ships from other planets but were of terrestrial origin and came from a subterranean race dwelling inside the earth.

At first, the author could not accept this strange, unorthodox theory concerning the origin of the flying saucers, which seemed improbable and impossible, since it would require the existence of a cavity of tremendous size inside the earth in which they could fly, in view of their tremendous speed. In fact, this cavity would have to be so large that it would make the earth a hollow sphere. At this time the author had not come across the remarkable books of two American scientists, William Reed and Marshall B. Gardner, proving, on basis of evidence from Arctic explorers, that the earth is hollow with openings at the Poles, with a diameter of 5,800 miles in its hollow interior, large enough for flying saucers to fly in.

Huguenin's theory of the subterranean origin of the flying saucers, however, was not original. The idea was first put forward by Professor Henrique Jose de Souza, president of the Brazilian Theosophical Society, which has its headquarters in Sao Lourenco in the State of Minas Gerais, where there is an immense temple in Greek style dedicated to "Agharta," the Buddhist name for the Subterranean World.

Among the professor's students at Sao Lourenco were Mr. Huguenin and Commander Paulo Justino Strauss, officer of the Brazilian Navy and member of the Diretoria of the Brazilian Theosophical Society. From him they learned about the Subterranean World, and also the idea that flying saucers come from the Earth's interior. It was for this reason that Mr. Huguenin dedicated his book to Prof. de Souza and his wife, D. Helena Jefferson de Souza.

While Huguenin incorporated the idea of the subterranean origin of the flying saucers in a book, Commander Strauss presented it in a series of lectures which he held in Rio de Janeiro, in which he affirmed that the flying saucers are of terrestrial origin, but do not come from any known nation on the earth's surface. They originate, he believes, in the Subterranean World, the World of Agharta, whose capital city is known as Shamballah.

In his book, Huguenin presents Strauss's views on the subterranean origin of the flying saucers and against the theory that they come from other planets as follows:

"The hypothesis of the extra-terrestrial origin of the flying saucers does not seem acceptable. Another possibility is that they are military aircraft belonging to some existing nation on earth. This hypothesis, however, is opposed by the following arguments:

"1. If the United States and Russia possessed flying saucers, they would not desist from announcing this fact because of its value as a psychological arm to secure advantages in the diplomatic field. Also they would manufacture and use these vehicles for military purposes, since they are so rapid and powerful that they would leave the enemy almost without means of defense.

"2. The United States and USSR would not continue to spend large sums of money on the manufacture of ordinary airplanes if they possessed the secret of producing flying saucers."

After presenting the argument that flying saucers do not come from any existing nation and his view that they are not of interplanetary origin, Huguenin quotes Strauss to the effect that they come from the Subterranean World. On this subject he writes:

"Finally, we must consider the most recent and interesting theory that has been offered to account for the origin of the flying saucers: the existence of a great Subterranean World with innumerable cities in which live millions of inhabitants. This other humanity must have reached a very high degree of civilization, economic organization and social, cultural and spiritual development, together with an extraordinary scientific progress, in comparison with whom the humanity that lives on the earth's surface may be considered as a race of barbarians.

"The idea of the existence of a Subterranean World will shock many people. To others it will sound absurd and impossible, for 'certainly,' they say, 'if it existed, it would have been discovered long ago.' And there are plenty of other critics who would point out that it would be impossible for such an inhabited world to exist inside the earth because of the belief that as one descends, the temperature increases, on the basis of which theory it is supposed that, since the temperature increased the further down one went, the center of the earth is a fiery mass. However, this increase in temperature does not mean that the center of the earth is fiery, since it might extend only for a limited distance and, as in the case of volcanos and hot springs, arise from subterranean cavities located at certain levels (below which the temperature again drops as one goes downward). In accordance with the hypothesis that heat increases as one descends through the earth's crust, this takes place only a distance of eighty kilometers (in the superficial layer of the earth).

"According to the information supplied by Commander Paulo Justino Strauss, the Subterranean World is not restricted to caverns, but is more or less extensive and located in a hollow inside the Earth large enough to contain cities and fields, where live human beings and animals, whose physical structure resembles those on the surface. Among its inhabitants are certain persons who came from the surface, who, like Colonel Fawcett and his son Jack, descended, never to return." (Huguenin here refers to the views of Professor de Souza and Commander Strauss on the controversial subject of Colonel Fawcett's mysterious disappearance, claiming that he and his son Jack are still living in a subterranean city to which they gained access through a tunnel in the Roncador Mountains of Northeast Matto Grosso, and were not killed by Indians as commonly supposed. Fawcett's wife, who claims to be in telepathic contact with him, is positive that he is still living, so much so that she sent an expedition to Matto Grosso, in charge of her other son, to find him, but in vain, because he was no longer on the earth's surface, but in the Subterranean World.)

Huguenin then asks how these marvelous subterranean cities and this advanced civilization in the interior of the earth arose. His answer is that the builders and most of the inhabitants of this Subterranean World are members of an antediluvian race which came from the prehistoric submerged continents of Lemuria and Atlantis, who found refuge there from the flood that destroyed their Motherland. (Lemuria sank under the Pacific Ocean 2,500 years ago, while Atlantis was submerged by a series of inundations, the last of which occurred 11,500 years ago, according to Plato's account, derived from ancient Egyptian records. Egypt was a colony of Atlantis to the East, just as the Aztec, Mayan and Inca empires were to the West.)

Huguenin claims that the Atlanteans, who were far in advance of us in scientific development, flew the sky in aircraft utilizing a form of energy obtained directly from the atmosphere, and which were known as "vimanas," which were identical with what we know as flying saucers. Prior to the catastrophe that destroyed Atlantis, the Atlanteans found refuge in the Subterranean World in the hollow interior of the earth, to which they traveled on their "vimanas" or flying saucers, reaching it through the polar openings. Ever since then, their flying saucers remained in the earth's interior atmosphere and were used for purposes of transportation from one point in the interior *concave world* to another, for in this world, inside the crust of the earth, a straight aerial line is the shortest distance between any two points, no matter how far apart. It was only after the Hiroshima atomic explosion that these Atlantean aircraft rose to the surface for the first time, and were known as flying saucers. As we have pointed out previously, they came as an act of self-defense, to prevent radioactive pollution of the air they receive from the outside.

Huguenin is convinced that flying saucers are not space ships from other planets, but Atlantean airships. It seems that throughout history, especially in ancient times, these aircraft occasionally rose to the surface, and some historical figures rode in them. Thus in the Indian

epic, "Ramayana," there is a description of a Celestial
Car of Rama, the great teacher of Vedic India, known
as "vimana," a controlled aerial vehicle. It was capable
of flying great distances. Rama's aerial record was a hop
from Ceylon to Mount Kailas in Tibet. In the "Maha-
bharata," we read of Chrishna's enemies having built an
aerial chariot with sides of iron and clad with wings. The
"Smranagana Sutrahara" says that by means of skyships
human beings can fly in the air and "heavenly beings"
would come down to earth.

That aerial navigation existed long before the making
of the first modern airplane by the Wright brothers, the
director of the International Academy of Sanskrit Investigation at Mysen, India, discovered in an ancient
treatise on aeronautics, which was written three thousand
years ago. It was attributed to the Indu sage Bharadway,
who wrote a manuscript called "Vymacrika Shostra,"
meaning "the Science of Aeronautics." It has eight
chapters with diagrams, describing three types of aircraft,
including apparatuses that could neither catch on fire
nor break, and mentions thirty-one essential parts of
these vehicles and sixteen materials from which they are
constructed, which absorb light and heat, for which
reason they were considered suitable for the construction
of airplanes. It is interesting to note the similarity of the
word "vymacrika" and "vimanas," indicating that the
Hindus obtained their knowledge of aerial navigation
from the subterranean Atlanteans who must have visited
them in ancient times and taught them.

From Brazil, where the theory of the subterranean
origin of the flying saucers originated, it spread to the

United States, where Ray Palmer, editor of "Flying Saucers" magazine became its enthusiastic proponent, abandoning his former belief in their interplanetary origin in favor of the new theory that they came from the hollow interior of the earth. In the December, 1959 issue of his magazine, he wrote:

"In this issue we have presented the results of years of research, in which we advance the possibility that the flying saucers not only are from our own planet, and not from space, inner or outer, but there is a tremendous mass of evidence to show that there is an UNKNOWN location of vast dimensions which is, insofar as we can safely state at this writing, also unexplored, where the flying saucers can, and most probably do originate."

In reference to the claims made by some flying saucer "contactees" that they were taken up on a flying saucer for a trip to Mars and other planets, Palmer says:

"We've read all the accounts of such voyages and nowhere, in any of them, can we find positive evidence that space was traversed! In all these accounts, we can see where the passengers could have been taken to this 'unknown land' discovered by Admiral Byrd, and if told they were on Mars, they would not know the difference!

"Provided an actual trip in a saucer was made, the pilots of the flying saucers could have simulated a space trip and instead took their passengers to 'that mysterious land beyond the Pole,' as Admiral Byrd calls it."

In an article, "Saucers From Earth: A Challenge to Secrecy," in the Dec. 1959 issue of "Flying Saucers," Palmer writes:

"Flying Saucers magazine has amassed a large file of

evidence which its editors consider unassailable, to prove that the flying saucers are native to the planet Earth: that the governments of more than one nation know this to be a fact; that a concerted effort is being made to learn all about them, and to explore their *native land*; that the facts already known are considered so important that they are the *world's top secret*; that the danger is so great that to offer public proof is to risk widespread panic; that public knowledge would bring public demand for action, which would topple governments both helpless and unwilling to comply; that the inherent nature of the flying saucers and their origination area [in the earth's hollow interior, reached through the polar openings—Author] is completely disruptive to political and economic status-quo."

As against the theory that flying saucers were made by any existing government, Palmer says, "Flying saucers have been with humanity for centuries, if not thousands of years." Their antiquity, he says, "eliminates contemporary earth governments as the originators of the mysterious phenomenon."

After disproving that flying saucers come from any existing nation, Palmer attacks the theory of their interplanetary origin, whose chief proponent is the American flying saucer expert, Keyhoe, also some "contactees" who claim some flying saucers come from Mars, others from Venus, etc.

After showing that flying saucers do not come from any existing nation or from other planets, Palmer, America's greatest authority on flying saucers, concludes, in agreement with Commander Strauss and Huguenin, that they come from the earth's hollow interior through the polar openings. He writes:

"In the opinion of the editors of 'Flying Saucers,' this Polar origin of the flying saucers will now have to be factually disproved. Any denial must be accompanied with positive proof. 'Flying Saucers' suggests that such proof cannot be provided. 'Flying Saucers' takes the stand that all flying saucer groups should study the matter from the hollow earth viewpoint, amass all confirmatory evidence available in the last two centuries, and search diligently for any contrary evidence. Now that we have tracked the flying saucers to the most logical origin (the one we have consistently *insisted* must exist because of the insurmountable obstacles of interstellar origin, which demands factors beyond our imagination), that the flying saucers come from our own Earth, it must be proved or disproved, one way or the other.

"Why? Because if the interior of the Earth is populated by a highly scientific and advanced race, we must make profitable contact with them; and if they are mighty in their science, which includes the science of war, we must not make enemies of them; and if it is the intent of our governments to regard the interior of the Earth as 'virgin territory,' and comparable to the 'Indian Territory' of North America when the settlers came over to take it away from its rightful owners, it is right for the people to know that intent, and to express their desire in the matter.

"The flying saucer has become the most important single fact in history. The momentous questions raised in this article are to be answered. Admiral Byrd has discovered a new and mysterious land, the 'center of the great unknown,' and the most important discovery of all time. We have it from his own lips, from a man whose integrity has always been unimpeachable, and whose mind was one of the most brilliant of modern times.

"Let those who wish to call him a liar step forward and prove their claim! Flying saucers come from this Earth!"

So ends Ray Palmer's great article, "Flying Saucers From the Earth!" which created a sensation, causing certain government secret agencies to confiscate the magazine and stop its distribution, so that it did not reach its 5000 subscribers. Why? Obviously because the government was convinced that such an unclaimed, unknown territory, vast in extent, larger than the entire land surface of the earth, exists and wished its existence to be kept secret, so that no other nation would know about it or reach it before and claim this territory as its own. It was important that the Russians not learn about it. For this reason it was decided to suppress this issue of "Flying Saucers" of December, 1959, which was mysteriously removed from circulation. Evidently the information contained in this magazine concerning the fact that flying saucers come from the earth's hollow interior through the polar openings, like news concerning Admiral Byrd's flights past the Poles into the new unknown territory beyond them, was considered dangerous to be released to the public and was consequently secretly suppressed by government authorities.

Another outstanding American authority on flying saucers is Gray Barker. A month after Palmer published his sensational article expressing his belief that flying saucers do not come from outer space but from the earth's interior, Barker, in his "The Saucerian Bulletin," on January 15, 1960, wrote:

"In the December 1959 issue of 'Flying Saucers' Ray Palmer came out with his findings. The theory had been advanced before, many years previously, in a book titled 'A Journey to the Earth's Interior, Or Have the Poles Really Been Discovered?', now out of print and very rare. Many occult students, long before flying saucers became widely known about, believed that People lived inside the earth, emerging and entering through secret openings at the North and South Poles.

"Palmer presented only the first of his evidences in the December 1959 issue. It consisted of a review of newspaper and radio accounts of Admiral Richard E. Byrd's flight to the North Pole in 1947.

"In February of that year, Byrd took off from an Arctic base and headed straight north to the Pole. Then Byrd kept flying north, beyond the Pole, and was amazed to discover iceless lands and lakes, mountains covered with trees, and even a monstrous animal moving through the underbrush below! For almost 1700 miles the plane flew over land, mountains, trees, lakes and rivers. After flying 1700 miles, he was forced to turn back because of his gasoline supply limit for the return trip. So he retraced the flight back to the Arctic base. Not much was thought about the unusual flight at the time.

"Palmer then instructs the reader to look at the globe. According to Byrd's reported flight, he shouldn't have seen anything but ice-covered ocean or partially open water. Yet Byrd saw trees and other greenery. According to the globe, such a land just isn't there.

"Palmer next discusses similar geographical discrepancies at the South Pole, and then draws the amazing conclusion:

"THE EARTH IS NOT SPHERICAL. INSTEAD IT IS SOMETHING LIKE A DOUGHNUT, though perhaps not so flattened. At each pole there is a huge opening, so large that when one travels 'beyond' the Pole, he actually enters the lip of the hole of the doughnut-shaped earth. If he traveled far enough he would travel through the 'hole' of the 'doughnut' and emerge at the other Pole.

"Palmer further suggests that people live on the 'inside' of the earth, and that such people emerge from the Poles in flying saucers. He promises to present the remainder of the proofs later, but in the present issue of 'Flying Saucer,' his case boils down to these main points:

"(1) Measurements of areas at the North and South Poles are larger than you can find room for on a map or globe, leading to the assumption that such areas extend down into the 'doughnut.'

"(2) Some animals, particularly the musk-ox, migrate north in the wintertime, from the Arctic Circle. Foxes are found north of the 80th parallel, heading north, and appear well fed in a large area where there is no food available. [They go north because it becomes warmer and there is plant and animal life as they enter the polar opening—Author.]

"(3) Arctic explorers agree it gets warmer as one heads north (after coming close enough to the North Pole).

"(4) In the Arctic, coniferous trees drift ashore, from out of the north. Butterflies and bees are found in the far north, but never hundreds of miles south of that point.

"(5) Remains of mammoths, perfectly preserved, were found in Siberia, with the sparse food of the sub-Arctic region in its stomach. Such food could not have supported the animal. It must have come from the 'land beyond the Poles,' Palmer postulates.

"(6) Trouble with satellites shot over the South Pole bears out the theory that land areas haven't been measured accurately or that 'somebody' has been interfering with them."

In this connection it is interesting to note that American newspapers, some time back, published a report of a mysterious artificial satellite discovered to encircle the earth in the orbit that passed directly over both Poles and which was sent by no known nation. Did it emerge from one of the Poles and continue to rotate around its point of origin?

Gray Barker seems to agree with Palmer that flying saucers come from inside the earth; and in his editorial quoted above, he asks: "What if there could be some unknown race, on some unexplored portion of the earth, which is responsible for the flying saucers? Palmer's articles started me to thinking along that direction once again. THE INNER EARTH EXPLANATION WOULD FIT INTO MOST, IF NOT ALL THE FACETS OF THE FLYING SAUCER PICTURE.

"Various occult schools teach that polar entrances provide the doorways to cities of Agharta, the Subterranean

World, such as Shamballah (the capital) and others. Let us accept, for a moment, that such a people has existed inside the earth for thousands of years, even before man—or maybe they seeded the outside with man. Maybe they have constantly watched over him, occasionally assisting him with technology, giving rise to what we now call 'legends.' Maybe they built the Great Pyramid; maybe they are responsible for some of the 'miracles' reported in secular and religious histories. Until man, their protege, learned to be morally worthy, they would not wish to give him, suddenly, the knowledge of their existence or secrets of their technology.

"When man, however, invented the atomic bomb, the people of the inner earth were greatly concerned about it. Maybe they feared that contamination of the atmossphere would reach them; maybe they feared man could blow up the earth entirely. Halting or controlling man's propensity for destruction would be a delicate problem unless they would come out openly and inform him of their existence. They figured that they would eventually have to do so, and began a slow process of indoctrination, first merely letting him see the flying saucers fly around. Since men thought that flying saucers came from outer space, they pretended to be space people contacting him in their craft, and trying to indoctrinate him with peaceful philosophy (the majority of 'space people' contacted having spoken strongly against the atomic bomb) ."

In his book, "They Knew Too Much About Flying Saucers," Barker speaks of the "Antarctic Mystery" or the unusual number of flying saucers seen to ascend and descend in the region of the South Pole, which strongly supports the theory of a polar opening through which flying saucers emerge from and enter the hollow interior of the earth. In this book he mentions an Australian and a New Zealand investigator, named Bender and Jarrold respectively, who believed that flying saucers originate and are based in the Antarctic and tried to trace their course, when they were suddenly stopped in their research by 'three men in black,' who were secret government agents who apparently wished to suppress such research, just as publicity concerning Admiral Byrd's 2,300 mile flight to the new unknown territory not found on any map, that lies beyond the South Pole and inside the opening that leads to the earth's hollow interior, was suppressed in the press.

Theodore Fitch is another American writer who believes that flying saucers come from the hollow interior of the earth. In his book, "Our Paradise Inside the Earth," he writes:

"Writers of books on flying saucers believe that they come from other planets. But how can that be? They are too far away. Traveling at terrific speeds it would take a lifetime to make the trip (especially from planets of other solar systems) ."

Fitch claims, as does Palmer, that the "spacemen" who come to us in flying saucers, who pose as visitors from other planets, are really members of an advanced civilization in the hollow interior of the earth, who have important reasons for keeping their true place of origin secret, for which reason they purposely foster the false belief that they come from other planets. On this point, Fitch writes:

"They say that they come from other planets, but we doubt it." He considers this a white lie in order to prevent militaristic governments from learning that on the opposite side of the earth's crust there exists an advanced civilization whose scientific attainments far surpass our own, which is reached by the polar openings. In this way they protect themselves from molestation or possible war between subterranean and surface races.

Fitch agrees with Palmer that flying saucers are not "space ships," as Adamski claims, nor are their pilots "spacemen." Rather they are vehicles for atmospheric travel which come from the hollow interior of the earth in which they fly, connecting each part of the concave subterranean world with the other. As for the "little brown men" seen in flying saucers, Fitch believes that they belong to the same subterranean race from which the Eskimos descended. Fitch is in agreement with William Reed and Marshall B. Gardner that the ancestors of the Eskimos came from the hollow interior of the Earth through the polar opening. Describing these little brown men, who are the pilots of the flying saucers, evidently serving a master race (Atlantean) which built them and sent them to us Fitch says:

"Though smaller than we, they are stronger. Their grip is like a vice. One of them could quickly overpower a strong man. Their bodies are perfect in build. Both men and women dress neatly. Though not beautiful, they are nice looking. Not one of them looks to be over 30 years old. They say that they do not expect to ever die.

"It would take a book to record the conversation that has taken place with the saucermen and women. Their speech is quick, sharp and right to the point. They seem to be very, very intelligent. They talk freely and answer all questions, but they lie about things they do not want us to know (refusing to reveal their true subterranean origin and pretending to come from other planets, as Mars and Venus).

"Here are a few brief statements or claims made by the little men and women who live inside the earth. They boast about their superior mentality and knowledge, and that they excel us in creative ability. They say they are far ahead of us from the standpoint of new inventions. For instance, they claim that their flying saucers are powered with 'free energy' (meaning the electromagnetic energy of space, which is free and not like fuel used to supply our aircraft). They claim they obtain this 'free energy' by exploding certain atoms by the action of the electromagnetic energy of space while in flight.

"They say they are thousands of years ahead of us in all of the arts, such as painting, sculpture and architectural designing. Also they are ahead of us in their domestic and business management, in their agricultural techniques; and their beautiful landscapes, parks, flower gardens, orchards and farms vastly surpass our own. They claim that they are far ahead of us in their knowledge of nutrition and diet.

"They claim to live in luxury, yet have no class distinction and no poverty among them, nor need of police. They say that they know every language on earth."

Fitch's description of this superior civilization in the hollow interior of the earth reminds one of Bulwer Lytton's subterranean Utopia described in his book, "The

Coming Race." Lytton was a Rosicrucian and probably had access to occult information along this line. He described a superior race inside the earth which lived in a state of universal abundance and contentment, free from greed, poverty and war.

Fitch describes these people as living under an economic system by which they own all things in common, without private aggrandizement or hoarding, and without class distinctions of rich and poor, capitalist or worker. Also they have an equitable system of distribution free from exploitation and usury; and there is no poverty among them, since all are on a basis of perfect equality through a system of common ownership. They have no private property and work together cooperatively for their mutual welfare. Fitch writes:

"They say they know all the secrets of every government. They say they are of higher intelligence and authority. Since they are our superiors they have authority over us. They claim to be experts in mental telepathy. They claim they came from an antediluvian race (Lemurian and Atlantean). They say they know nothing at all about our Jesus, and say our Bible has been mistranslated, misinterpreted and misconstrued. They claim that they are a race which has not fallen as we have. They say we should have a world government. They say we should get rid of nuclear bombs and armaments.

"They say all their efforts are for peace. They say our peace is due to their efforts in our behalf, and that they saved us from being plunged into a suicidal nuclear war, and that we should look up to them for guidance.

"Pictures have been taken of the little brown saucermen and their words are recorded on tape. Certain Americans have taken both short and long rides in both small and large flying saucers."

We have shown above that flying saucers are atmospheric vehicles created by a super race living in the hollow interior of the earth and are not space ships that came from other planets, as is commonly supposed without a particle of evidence in its favor. Instead we have positive proof that this cannot be so.

The fact that the mass visitation of flying saucers occurred following the explosion of the first atomic bomb in Hiroshima has been supposed by some writers to indicate that the flash of the explosion attracted the attention of inhabitants of other planets or solar systems, who sent their flying saucers to us to prevent a catastrophe that might endanger the universe, themselves included. For this reason, it is claimed, the mass visitation took place after the Hiroshima disaster, whereas previously, flying saucers appeared only spasmodically and never in such great number.

This idea is senseless for several reasons. First of all, granted that flying saucers come from other planets and solar systems, some many light years away, and requiring twice as long a time for the flash to reach them and for them to come to Earth (provided that they could travel at the speed of light), how could it be possible for flying saucers from different planets and solar systems to all arrive here at approximately the same time, and so soon after the Hiroshima explosion? This alone should disprove the theory of the interplanetary origin of the flying saucers.

Secondly, if their coming was an act of self-defense,
lest the first atomic explosion later lead to a much
greater release of atomic energy, whose dangerous effect
will be to poison our atmosphere, it is much more reasonable to believe that subterranean inhabitants, who derive
the air they breathe from the outside (through the polar
openings) would be most fearful of such a calamity and
send their fleets of flying saucers in order to befriend us,
gain our respect and then counsel us to desist from further atomic explosions and manufacture of atomic
bombs. Certain inhabitants of other solar systems many
light years away would have no cause to be unduly concerned about our poisoning of our atmosphere or even
if the Earth exploded and was transformed into meteors.
And if they traveled so far to prevent further nuclear explosions or the production of more destructive bombs,
their voyage was in vain. Also, since flying saucers, after
1945, came in greater number, and for purposes of observation, emissaries from other solar systems could
achieve their purpose by sending a single unit and have
no need to send a fleet. They would have less reason to
be concerned with a wayward planet many light years
or millions of miles away than earth dwellers who live in
its interior, who must suffer from any radioactive contamination of the air they receive from outside.

Since the purpose of the coming of the flying saucers
was to prevent a radioactive contamination of the atmosphere and the destruction of the human race through
a nuclear war (perhaps they did prevent a war that might
have occurred had they not come to help us), by making
it known to the heads of our governments that a super

race exists possessing scientific powers far in advance of our own, and hoping to gain their respect so that they will obey their admonitions to desist from further playing with atomic fire, this will explain why they came in fleet formation to attract public attention and also appeared so often near military airports to convince Air Force chiefs of their existence, feeling that their reports would have most weight with the Government. Once their existence was recognized, they hoped to convince the U.S. Government, and through it, all governments to desist from further atomic experimentation and the production of nuclear bombs.

But their plan to save humanity (and themselves as well) failed. Instead of recognizing and admitting their existence, as a superior race that came to enlighten and help us, and prevent us from committing nuclear suicide, and in spite of undeniable evidence of their existence in the possession of the U.S. Air Force, Government leaders refused to believe that they are real, and since they were not believed to exist, no effort was made to cooperate with their plan to avert a world catastrophe and the radioactive destruction of the human race (now in progress in the form of radioactive fallout which has reached the danger point in the Northern Hemisphere according to recent measurements by an Italian scientist in Rome).

Rather than show reverence for these superior beings possessing a scientific development far beyond our own, as shown by the superiority of their aircraft (flying saucers) over ours, instead of receiving them in a friendly way, whenever a flying saucer came near a United States military airfield, pursuing planes were sent after them, with instructions to open fire on them and down them, in order to discover the secret of their construction and source of power. In the famous "Captain Mandell" incident, he pursued a flying saucer which appeared near a military airport as it rose higher and higher until his plane mysteriously exploded.

Disappointed with their efforts to befriend and establish friendly contact with surface humanity, the chiefs of the flying saucer fleets that appeared in our skies after 1945, continuing in numbers for some years after, later withdrew the large numbers of flying saucers they formerly sent to us, when they had hopes to befriend us and so convince us to desist from further atomic explosions and experimentation, and from further manufacture of atomic bombs. The number of flying saucers left in our atmosphere was few, as is the case today. The few that were left probably are acting as scouts to conduct measurements of radioactive fallout and atmospheric contamination, which they communicate to scientists in subterranean headquarters.

But there are many other arguments against the interplanetary hypothesis of the origin of the flying saucers. This theory does not explain how, under entirely different geological, chemical, atmospheric, gravitational, climatic and other conditions, planets millions or trillions of miles away, and belonging to other solar systems, could develop human beings so like us in structure, appearance, clothing, customs, language, accent and ideas as the "Venusians" whom Adamski claimed to have met in a "mother ship" or "space ship" he claims to have visited.

The fact that these people not only look like us, have the same stature and even speak with an accent (in many cases, a German accent), seems strange if they came from another planet. It seems much more probable that they came originally from the earth's surface, gained access to the Subterranean World, and are employed as pilots by subterranean authorities, who sent them to us.

If they came from other planets or solar systems, it would be very improbable that they would look and speak like us as much as they do. Most writers of science fiction imagine inhabitants of other planets to be entirely unlike us in their structure. In his "War of the Worlds," H. G. Wells pictured Martians as mechanical monsters. It would be a rare coincidence that other planets would develop forms of life so much like our own as are the pilots of the flying saucers, according to those who claim to have met them. As for the "small men" found in flying saucers, they are probably subterranean dwarfs employed by the master race that created them as pilots.

If the people seen in flying saucers were members of our own race (chiefly Germans, since so many of them speak German, which would be strange if they came from other solar systems or planets), employed as pilots, they would probably have been instructed by their leaders not to reveal the secret of the origin of the flying saucers for the reason that the land area of the New World in the hollow interior of the earth is greater than that of its surface, where we find more area covered by ocean water, and should militaristic governments learn about this, they would make a mad rush to send their aircraft through the polar openings to claim this territory as their own, just as the governments of Europe sent their expeditions to America soon after Columbus discovered the new continent.

If certain ambitious surface governments sought to appropriate this new territory, enjoying an ideal subtropical climate, by force, by sending in expeditions equipped with nuclear weapons, the superior subterranean people would be forced to defend themselves by their "death rays," a force far more powerful than atomic energy, capable of bringing about complete atomic disintegration and the dematerialization and disappearance of their invaders and their weapons. Such a catastrophe they would rather prevent, since they are pacifists and detest warfare.

For this reason they wished to keep the existence of the subterranean world a secret, so that its inhabitants may not be molested by invaders from the outside. For this reason flying saucer pilots were instructed to pretend that they came from other planets and were "spacemen," in case they were contacted, and to keep it a secret that they came from inside the earth. In this way they could guard their secret. Adamski and others who claimed to contact them were accordingly deceived by the false idea that the flying saucer travelers came from other planets.

If the major governments would forget their race into space and send armies of ice breakers, dirigibles and aircraft to penetrate as far as possible through the polar openings, it would not be long before contact would be established between the superior race living inside the earth's crust and the less advanced race, still in a state of

mechanized barbarism and engaging in constant wars,
inhabiting the earth's surface. However, militaristic governments are not worthy of establishing contact with

WATER-SKY

Here is another view showing how the surface of the
earth, water, and ice is reflected as in a mirror.

such superior, superhuman beings, who would probably use their powerful radiations capable of dematerialization to prevent intrusion by undesirable or dangerous visitors. Since they came from Atlantis, which had a civilization far greater than our own over 11,500 years ago and for many thousands of years previously, this elder race has a scientific development as much greater than our own as ours is greater than that of the Hottentots.

In comparison with the superior Subterranean People, surface dwellers are barbarians, and their proud "civilization" is a state of mechanized barbarism. Until they learn to relinquish war forever, to destroy and bury all nuclear weapons, to establish a world government, a world court and a world police, and until they reorganize their economic and financial system on a basis of equity and justice, they will be unworthy to contact the inhabitants of the Subterranean World, who stand on a level of scientific, intellectual and moral development vastly beyond those on the surface.

Chapter VIII

DESCRIPTION OF A THEORETICAL AERIAL EXPEDITION INTO THE POLAR OPENING LEADING TO THE HOLLOW INTERIOR OF THE EARTH

Marshall B. Gardner ends his great book by describing a theoretical expedition as it approached the polar opening, entered into it and then reached the tropical paradise in the hollow interior of the earth. His object was to encourage some government to undertake such an expedition. Admiral Byrd was the first one who did so. But he did not enter far enough to reach the Subterranean World. He reached only its periphery.

On September 15, 1959, the Soviet atomic powered icebreaker was first launched and was supposedly on its way to the North Pole with intentions of reaching it by smashing all its way through the ice. "What better mode of travel can there be to proceed into that 'unknown land' beyond the Pole, which extends uncounted thousands of miles?" asked Ray Palmer, who adds:

"Here we have a craft that has a cruising range of 40,000 miles. It can go anywhere with no danger of being stranded for lack of fuel. It is a craft exactly suited for bridging the ice barrier of a frozen ocean that has always been the 'wall' between the known world and the 'unknown' world that Admiral Byrd proved beyond doubt to exist.

"Once through the ocean of ice, into a warm ocean, it is admirably suited for exploration deep into that unknown area, as far as water exists. It may be that the Russians are unaware of Admiral Byrd's discovery and the icebreaker will not go 'beyond' the Pole."

On January 13, 1956, a United States Navy air expedition, commanded by Admiral Byrd, flew for 2,700 miles from its base at McMurdo Sound, which is 400 miles west of the South Pole and penetrated a land extent of 2,300 miles beyond the Pole into the South Polar Opening leading to the hollow interior of the earth. This was the first time in history that members of the humanity dwelling on the surface of the earth had penetrated so far into the earth's interior. If the expedition had traveled a few thousand miles more, it would have reached the great civilization that exists inside the earth, which has sent its flying saucers to us—a civilization which is thousands of years in advance of our own in its scientific achievements, moral perfection and social, economic and political organization. Thousands of years ago it had established a state of permanent peace under a world government, and had abolished the evil of war. While the civilization on the surface was continually interrupted in its progress and suffered constant retrogressions as a result of never-ending wars, the inhabitants of the earth's interior, who were free from this impediment, made continual scientific progress, as is evident by their scientific superiority over us in the art of aerial navigation —their *flying saucers*. To contact such a highly evolved race would indeed be a great privilege and one of the greatest discoveries ever made in human history. It will depend on some courageous aviator or air expedition to make this discovery, a discovery far greater than Columbus's discovery of America.

Let us now describe a theoretical journey through the South Polar opening to the New World that lies beyond. The best air vehicle for this purpose would be a dirigible (zeppelin), which has many advantages over the airplane. In case it ran out of fuel in this long voyage, it could radio for aid and not risk danger of crashing to the ground.

The first stop in such an expedition heading to the South Pole would be Tierra del Fuego at the extreme south of South America, not far from the continent of Antarctica. Here the gasoline supply could be replenished. Then the expedition would travel straight south, and after passing 90 degrees south latitude, it would proceed in the same direction, regardless of the eccentricities of the compass. In time it would leave the barren waste of Antarctic ice and enter a territory of flora and fauna, as Admiral Byrd did when he traveled past the North Pole for 1,700 miles. The expedition could then photograph the vegetable and animal life in this Land Beyond the Pole by flying low enough.

As the expedition advances into the polar opening, after the setting of the sun, there will be observed a glow in the sky which appears like a ring covering the visible horizon, formed by the aurora, which appears as long streamers of light which wave in fantastic patterns. These lights result from the reflection of the central sun on the higher strata of the atmosphere, which is illuminated for an immense area by its diverging rays. As the expedition proceeds, the auroral displays become brighter and brighter.

As the expedition advances deeper and deeper into the polar opening, the sun gets nearer and nearer to the horizon each day, and rises lower in the sky than it formerly did. It rises later and sets later. This is due to the rays being cut off by the rim of the polar aperture as the expedition enters it. Finally a strange thing happens. It is daylight when it should be night. Only it is a different daylight than we are accustomed to on the earth's surface —the sun being much dimmer and more reddish—for it is no longer the sun to which we are accustomed—the outer sun—but an inner sun which never sets and shines continually, producing perpetual daylight. Meanwhile, the temperature gets warmer and warmer, until the climate becomes tropical, a climate of perpetual summer without changes of seasons.

As the expedition proceeds, it will notice that the sun now visible is no longer moving but is stationary in the sky. Finally it will observe new strange forms of tropical plant and animal life, including prehistoric species now extinct on the surface. This will be a veritable paradise for the botanist and zoologist.

Finally the expedition will pass the polar opening and reach the hollow interior of the earth—its interior atmosphere, the native home of the flying saucers. In time the expedition will commence to see signs of civilization and the subterranean cities of the Atlanteans and Lemurians who colonized this world many thousands of years ago, the creators of the flying saucers. The members of the expedition will then land and contact these highly civil-

ized people who will have much to teach them that will
be of utmost value to the human race. The message that
they will deliver will most probably relate to saving humanity from nuclear annihilation. Perhaps these people
hope to prevent the coming of World War III in
the near future. Or perhaps they are concerned with saving a remnant of the human race in the event that the
rest of humanity is exterminated, and colonizing them in
their Subterranean World, so that the human race might
not be entirely destroyed. These Atlanteans should have
much sympathy for us because their civilization also was
destroyed by a nuclear war followed by a flood, from
which they saved themselves in time by finding refuge

A bird's-eye view of the opening to the interior of the
earth.

in the Subterranean World. Since they foresee the same danger to us, they would probably like to save us in the same way they saved themselves when the rest of their countrymen perished.

Therefore, the members of this expedition may accomplish a mission of the utmost importance to the human race and may be hailed in the future not merely as the greatest explorers in history but as true Nuclear Age Saviors.

Chapter IX

AGHARTA, THE SUBTERRANEAN WORLD

The word "Agharta" is of Buddhist origin. It refers to the Subterranean World or Empire in whose existence all true Buddhists fervently believe. They also believe that this Subterranean World has millions of inhabitants and many cities, all under the supreme domination of the subterranean world capital, Shamballah, where dwells the Supreme Ruler of this Empire, known in the Orient as the King of the World. It is believed that he gave his orders to the Dalai Lama of Tibet, who was his terrestrial representative, his messages being transmitted through certain secret tunnels connecting the Subterranean World with Tibet. Similar mysterious tunnels honeycomb Brazil. Brazil in the West and Tibet in the East seem to be the two parts of the Earth where contact between the Subterranean World and the surface world may be most easily achieved, due to the existence of these tunnels.

The famous Russian artist, philosopher and explorer, Nicholas Roerich, who traveled extensively in the Far East, claimed that Lhasa, capital of Tibet, was connected by a tunnel with Shamballah, capital of the subterranean empire of Agharta. The entrance of this tunnel was guarded by lamas who were sworn to keep its actual whereabouts a secret from outsiders, by order of the

Dalai Lama. A similar tunnel was believed to connect the secret chambers at the base of the Pyramid of Gizeh with the Subterranean World, by which the Pharaohs established contact with the gods or supermen of the underworld.

The various gigantic statues of early Egyptian gods and kings, as those of Buddha found throughout the Orient, represent subterranean supermen who came to the surface to help the human race. They are generally represented as sexless. They were emissaries of Agharta, the subterranean paradise which it is the goal of all true Buddhists to reach.

Buddhist traditions state that Agharta was first colonized many thousands of years ago when a holy man led a tribe which disappeared underground. The gypsies are supposed to come from Agharta, which explains their restlessness on the Earth's surface and their continual travels to regain their lost home. This reminds one of Noah, who was really an Atlantean, who saved a worthy group prior to the coming of the flood that submerged Atlantis. It is believed that he brought his group to the high plateau of Brazil where they settled in subterranean cities, connected with the surface by tunnels, in order to escape from poisoning by the radioactive fallout produced by the nuclear war the Atlanteans fought, which brought on the flood that submerged their continent.

The subterranean civilization of Agharta is believed to represent a continuation of Atlantean civilization, which, having learned the lesson of the futility of war, has remained in a state of peace ever since, making stupendous

scientific progress uninterrupted by the setbacks of re-
current wars, as our surface civilization has been. Their
civilization is many thousands of years old (Atlantis sank
about 11,500 years ago), while ours is very young, only
a few centuries old.

Subterranean scientists are able to wield forces of nature we know nothing about, as demonstrated by their
flying saucers, which are operated by a new, unknown
source of energy, more subtle than atomic energy. Ossendowski claims that the Empire of Agharta consists of
a network of subterranean cities connected with each
other by tunnels through which vehicles pass at tremendous speed, both under land and under the ocean.

These people live under the benign reign of a world
government headed by the King of the World. They
represent descendants of the lost continents of Lemuria
and Atlantis, as well as the original perfect race of Hy-
perboreans, the race of gods.

During various epochs in history, the Aghartan supermen or gods came to the surface to teach the human race
and save it from wars, catastrophes and destruction. The
coming of the flying saucers soon after the first atomic
explosion in Hiroshima represents another such visitation, but this time the gods themselves did not appear
among men, but they sent their emissaries.

The Indian epic, "Ramayana" describes Rama as such
an emissary from Agharta coming on an aerial vehicle,
which was probably a flying saucer. A Chinese tradition
speaks of divine teachers coming on aerial vehicles.
Similarly, the founder of the Inca dynasty, Manco
Copac, came the same way.

One of the greatest of Aghartan teachers in America was Quetzalcoatl, the great prophet of the Mayas and Aztecs and of the Indians of the Americas in general, both in South and North America. That he was a stranger among them, coming from a different race (Atlantean) is indicated by his being fair, while they were dark; his being tall, while they were short; his being bearded, while they were beardless. He was reverenced as a savior by the Indians of Mexico, Yucatan and Guatemala long before the coming of the white man. The Aztecs called him "God of Abundance" and the "Morning Star." His name Quetzalcoatl means "Feathered Serpent," meaning a teacher of wisdom (symbolized by the serpent) who flies. He was given this name because he came on an aerial vehicle, which appears to have been a flying saucer. He probably came from the Subterranean World, because after he remained some time with the Indians, he mysteriously vanished the same way as he came; and was believed to have returned to the Subterranean World from which he came.

Quetzalcoatl is described as having been "a man of good appearance and grave countenance, with a white skin and beard, and dressed in a long flowing white garment. He was also called Huemac, because of his great goodness and continence. He taught the Indians the way of virtue and tried to save them from vice by giving them laws and counsel to restrain them from lust and to practice chastity. He taught pacifism and condemned violence in all forms. He instituted a vegetarian diet, with corn as a principal food, and taught fasting and body hygiene. According to the South American archeologist, Harold Wilkins, Quetzalcoatl was also the spiritual teacher of the ancient inhabitants of Brazil.

After remaining some time with the Indians, and seeing how little they cared to follow his teachings, except his recommendation to plant and eat corn as a basic food in place of meat, Quetzalcoatl departed, telling them that some day he would return. That this "visitor from Heaven" left the same way in which he came—on a flying saucer—is indicated by the following facts. When Cortez invaded Mexico, the emperor Montezuma believed that the predicted "return of Quetzalcoatl" had occurred, because a fireball then gyrated over Mexico City, making the people wail and scream, setting the temple of the war god on fire. This fireball was believed to have been the flying saucer on which Quetzalcoatl traveled.

Osiris was another such subterranean god. According to Donnelly, in his book, "Atlantis the Antediluvian World," the gods of the ancients were the rulers of Atlantis and members of a superhuman race which governed the human race. Before the destruction of their continent, which they foresaw, they traveled by flying saucer through the polar opening to the Subterranean World in the hollow interior of the earth, where they have continued to live ever since.

"The Empire of Agharta," wrote Ossendowski in his book "Beasts, Men and Gods," "extends through subterranean tunnels to all parts of the world." In this book he speaks of a vast network of tunnels constructed by a prehistoric race of remotest antiquity, which passed under both oceans and continents, through which swift-

moving vehicles traveled. The empire of which Ossendowski speaks, and concerning which he learned
from lamas in the Far East, during his travels in Mongolia, obviously consists of subterranean cities inside the
earth's crust, which should be differentiated from those
existing in its hollow center. Thus there are two subterranean worlds, one more superficial and one in the
center of the earth.

Huguenin, whose book on flying saucers and the subterranean world we previously mentioned, believes that
there exist many subterranean cities at various depths,
between the earth's crust and its hollow interior. Concerning the inhabitants of these subterranean cities, he
writes:

"This other humanity has reached an elevated grade
of civilization, economic and social organization and
cultural and scientific progress, in comparison with which
the humanity which lives on the earth's surface are a race
of barbarians." In his book, Huguenin shows a diagram
of the earth's interior, showing various subterranean cities
at various depths, connected with each other by tunnels.
He describes these cities as existing in immense cavities
in the earth. The city of Shamballah, the capital of the
subterranean empire, he portrays as existing at the center
of the earth, in its hollow interior, rather than inside its
solid crust. Ossendowski writes:

"All the subterranean caverns of America are inhabited by an ancient people who disappeared from the
world. These people and the subterranean regions where
they dwell are under the supreme authority of the King
of the World. Both the Atlantic and Pacific Oceans were

once the home of vast continents which later became submerged; and their inhabitants found refuge in the Subterranean World. The profounder caverns are illuminated by a resplendent light which permits the growing of cereals and other vegetables, and gives the inhabitants a long life-span free from disease. In this world exists a large population and many tribes."

In his book, "The Coming Race," Bulwer Lytton describes a subterranean civilization far in advance of our own, which existed in a large cavity in the earth, connected with the surface by a tunnel. This immense cavity was illuminated by a strange light which did not require lamps to produce it, but appeared to result from an electrification of the atmosphere. This light supported plant life and enabled the subterranean people to grow their foods. The inhabitants of the Utopia described by Lytton were vegetarians. They had certain apparatuses by which, instead of walking, they flew. They were free from disease and had a perfect social organization so that each received what he needed, without exploitation of one by another.

It is claimed that the earth's crust is honeycombed by a network of tunnels passing under the ocean from continent to continent and leading to subterranean cities in large cavities in the earth. These tunnels are especially abundant in South America, especially under Brazil, which was the chief center of Atlantean colonization; and we may believe they were constructed by the Atlanteans. Most famous of these tunnels is the "Roadway of the Incas" which stretches for several hundred miles south of Lima, Peru, and passes under Cuzco, Tia-

huanaco and the Three Peaks, proceeding to the Ata-
cambo Desert. Another branch opens in Arica, Chile,
visited by Madame Blavatsky.

It is claimed that the Incas used these tunnels to
escape from the Spanish conquerors and the Inquisition,
when entire armies entered them, carrying with them
their gold and treasures on the backs of llamas, which
they did when the Spanish Conquerors first came. Their
mysterious disappearance at this time, leaving only the
race of Quechua Indians behind, is also explained by
their entering these tunnels. It is claimed that when
Atahualpa, the last of the Inca kings, was brutally
murdered by Pizarro, the gold that was being carried to
his ransom on a train of 11,000 pack llamas, found
refuge in these tunnels. It is claimed that these tunnels
had a form of artificial lighting and were built by the
race that had constructed Tiahuanco long before the
first Inca appeared in Peru. Since the Incas who entered
these tunnels to escape from the Spaniards were never
seen since and disappeared from the earth's surface, it is
probable that they continued to live in illuminated subterranean cities to which these tunnels led.

These mysterious tunnels, an enigma to archeologists,
exist in greatest number under Brazil, where they open
on the surface in various places. The most famous is in
the Roncador Mountains of northeast Matto Grosso to
where Colonel Fawcett was heading when last seen. It
is claimed that the Atlantean city for which he searched
was not the ruins of a dead city on the surface but a
subterranean city with still living Atlanteans as its inhabitants; and that he and his son Jack reached this city

and are still living therein. This is the belief of Professor de Souza, Commander Strauss and O. C. Huguenin, whom we have mentioned before.

The Roncador tunnel opening is guarded by fierce Chavantes Indians who kill anyone who dares to enter uninvited and who might molest the subterranean dwellers whom they respect and reverence. The Murcego Indians also guard these secret tunnel openings leading to subterranean cities in the Roncador Mountain region of Matto Grosso. We quote a letter to the author from an American, named Carl Huni, who lived many years in Matto Grosso and made a special study of this subject:

"The entrance to the caverns is guarded by Murcego Indians, who are a dark-skinned, undersized race of great physical strength. Their sense of smell is more developed than that of the best bloodhounds. Even if they approve of you and let you enter the caverns, I am afraid that you will be lost to the present world, because they guard the secret very carefully and may not let those who enter leave. (This may have happened to Colonel Fawcett and his son Jack, who are believed to have entered a tunnel leading to a subterranean city in the Roncador Mountains, never to return.)

"The Murcego Indians live in caverns and go out at night into the surrounding jungles, but they have no contact with the subterranean dwellers below, inhabiting a subterranean city in which they form a self-contained community and have a considerable population. It is believed that the subterranean cities they inhabit were first constructed by the Atlanteans. One thing is certain, that no radioactive fallout can reach them. No one knows whether those who live in these ancient Atlantean sub-terranean cities are Atlanteans themselves or others who settled there after their original builders were gone. The name of the mountain range where these Atlantean subterranean cities exist is Roncador in northeast Matto Grosso. If you go in quest of these subterranean cities, take your life in your own hands as you may never be heard of again, like Colonel Fawcett.

"When I was in Brazil I heard a lot about the under-ground caverns and subterranean cities. They are, how-ever, a long way from Cuiaba. They are near the Rio Araguaya, which empties into the Amazon. They are to the northeast of Cuiaba at the foot of the tremendously long mountain range named Roncador. I desisted to investigate further because I heard that the Murcego Indians jealously guard the entrance to the tunnels from people who are not sufficiently developed, because they do not want trouble. In the first place, they do not want anyone who is still enmeshed in commercialism and who has a desire for money.

"I know that a good part of the immigrants who helped in the uprising of General Isidro Lopez back in 1028 disappeared into these mountains and were never seen again. It was under the reign of Dr. Benavides who bombarded Sao Paulo for four weeks. Finally they made a truce for three days and let the 4000 troops, who were mainly Germans and Hungarians, to go out of town. About 3000 of them went to Acre in the northwestern part of Brazil and about 1000 disappeared in the caverns. I heard the story consistently. If I remember the place where they disappeared was in the southern end of Bananal Island (near Roncador Mountains).

"There are also caverns in Asia and Tibetan travelers mention them. But as far as I know, in Brazil are the biggest ones and they exist at three different levels. I am sure I would get permission if I wanted to join them and they would accept me as one of theirs. I know they use no money at all, and their society is organized on a strictly democratic basis. People do not become aged and live in everlasting harmony."

This subterranean Utopia mentioned by Mr. Huni (now residing in New York) seems to resemble greatly the one described by Bulwer Lytton in his book, "The Coming Race." Lytton was a Rosicrucian and probably based his novel on occult information concerning existing subterranean cities.

The ruins of a number of Atlantean cities were found in northern Matto Grosso and the Amazon territory, indicating that Atlanteans once colonized this country. Some years ago an English schoolteacher, hearing rumors of a lost Atlantean city on a high plateau in this region went to find it. He did, but the hardships of the journey cost his life. Before he died he sent by carrier pigeon a note describing a magnificent city he discovered whose streets were lined by high gold statues.

If the Atlanteans once colonized Brazil and constructed cities in Matto Grosso on its surface, why did they build subterranean cities there? It could not have been to escape the deluge that submerged Atlantis and outlying areas, because Matto Grosso is a high plateau where floodwaters could not have reached. The South American archeologist, Harold Wilkins, offers another theory: that *the subterranean cities were built to escape the radioactive fallout resulting from a nuclear war the Atlanteans fought*. This seems to be a very reasonable explanation, for otherwise there would be no reason to undergo the great labor of excavating the earth and constructing subterranean cities when the Atlanteans already had magnificent cities on the earth's surface.

If and when we are endangered by a nuclear war, we, too, will have to find refuge inside the earth and dwell there in illuminated subterranean cities and produce our foods under this light. It would of course be much easier to join existing subterranean cities constructed by the Atlanteans thousands of years ago, who vastly surpassed us in engineering skill, than to construct our own. If friendly contact with subterranean dwellers could be established, when war came, or even before, when radioactive fallout increases beyond the danger point and menaces our survival, it would be to our advantage to contact these subterranean cities and, if we are admitted, to establish residence in them.

There is no old age in Agharta and no death. It is a society in which everyone is young looking, even if many centuries or even thousands of years of age. This seems incredible to surface dwellers exposed to the harmful effects of solar radiation and the autointoxication of food poisoning from a wrong diet. The symptoms of old age are not the natural result of the passage of time or an assumed aging process, but of adverse biological conditions and habits. Senility is a disease; and since

Aghartans are free from disease, they do not grow old.

The sexes live apart in Agharta and marriage does not exist. Each is free and independent and one sex does not depend on the other for its economic support. Reproduction is by parthenogenesis; and the virgin-born children are all female. (In this matriarchal civilization the female is considered the normal, perfect and superior sex.) Children are raised collectively by special teachers and not by private families. They are supported by the community. So are their mothers.

The superior scientific culture of the subterranean people, of which their flying saucers are an evident example, is the result of superior brain development and more energetic brains. This is due to the fact that their vital energies flow up to their brain, rather than being dissipated through the sexual channel as among so-called "civilized" surface races. In fact, sex indulgence is completely out of their lives; because of their fruit diet, their endocrines are in a state of perfect balance and harmonious functioning, as in little children, and are not stimulated to abnormal activity by metabolic toxins, as produced by such foods as meat, fowl, fish and eggs and by such aphrodisiacs as salt, pepper, coffee, tobacco and alcohol. By keeping their blood-stream pure and free from toxins, the subterranean people are able to live in complete continence, conserving all vital energies and converting them into superior brain power. Their superior scientific achievements result from the fact that their brains are superior to ours in intellectual development. They are the race which created the flying saucers.

Concerning Agharta, Professor Henrique J. de Souza,

President of the Brazilian Theosophical Society and a leading authority on the Subterranean World, in his magazine published an article he wrote, "Does Shangri-la Exist?" from which we quote:

"Among all races of mankind, back to the dawn of time, there existed a tradition concerning the existence of a Sacred Land or Terrestrial Paradise, where the highest ideals of humanity were living realities. This concept is found in the most ancient writings and traditions of the peoples of Europe, Asia Minor, China, India, Egypt and the Americas. This Sacred Land, it is said, can be known only to persons who are worthy, pure and innocent, for which reason it constitutes the central theme of the dreams of childhood.

"The road that leads to this Blessed Land, this Invisible World, this Esoteric and Occult Domain, constitutes the central quest and master key of all mystery teachings and systems of initiation in the past, present and future. This magic key is the 'Open Sesame' that unlocks the door to a new and marvelous world. The old Rosicrucians designated it by the French word VITRIOL, which is a combination of the first letters of the sentence: 'VISTA INTERIORA TERRAE RECTIFICANDO INVENES OMNIA LAPIDEM,' to indicate that 'in the interior of the earth is hidden the true MYSTERY.' The path that leads to this Hidden World is the Way of Initiation.

"In ancient Greece, in the Mysteries of Delphos and Eleusis, this Heavenly Land was referred to as Mount Olympus and the Elysian Fields. Also in the earliest Vedic times, it was called by various names, such as Ratnasanu (peak of the precious stone), Hermadri (mountain of gold) and Mount Meru (home of the gods and Olympus of the Hindus). Symbolically, the peak of this sacred mountain is in the sky, its middle portion on the earth and its base in the Subterranean World.

"The Scandinavian Eddas also mention this celestial city, which was in the subterranean land of Asar of the peoples of Mesopotamia. It was the Land of Amenti of the Sacred Book of the Dead of the ancient Egyptians. It was the city of Seven Petals of Vishnu, and the City of the Seven Kings of Edom or Eden of Judaic tradition. In other words, it was the Terrestrial Paradise.

"In all Asia Minor, not only in the past but also today, there exists a belief in the existence of a City of Mystery full of marvels, which is known as SHAMBALLAH (Shamb-Allah), where is the Temple of the Gods. It is also the Erdami of the Tibetans and Mongols.

"The Persians call it Alberdi or Aryana, land of their ancestors. The Hebrews called it Canaan and the Mexicans Tula or Tolan, while the Aztecs called it Maya-Pan. The Spanish Conquerors who came to America believed in the existence of such a city and organized many expeditions to find it, calling it El Dorado, or City of Gold. They probably learned about it from the aborigines who called it by the name of Manoa or City Whose King Wears Clothing of Gold.

"By the Celts, this holy land was known as 'Land of the Mysteries'—Duat or Dananda. A Chinese tradition speaks of Land of Chivin or the City of a Dozen Serpents. It is the Subterranean World, which lies at the roots of heaven. It is the Land of Calcas, Calcis or

Kalki, the famous Colchida for which the Argonauts sought when they set out in search of the Golden Fleece.

"In the Middle Ages, it was referred to as the Isle of Avalon, where the Knights of the Round Table, under the leadership of King Arthur and under the guidance of the Magician Merlin, went in search of the Holy Grail, symbol of obedience, justice and immortality. When King Arthur was seriously wounded in a battle, he requested his companion Belvedere to depart on a boat to the confines of the earth, with the following words: 'Farewell, my friend and companion Belvedere, and to the land where it never rains, where there is no sickness and where nobody dies.' This is the Land of Immortality or Agharta, the Subterranean World. This land is the Walhalla of the Germans, the Monte Salvat of the Knights of the Holy Grail, the Utopia of Thomas More, the City of the Sun of Campanella, the Shangri-la of Tibet and the Agharta of the Buddhist world."

We have indicated previously that the subterranean cities of Agharta were constructed by Atlanteans as refuges from the radioactive fallout produced by the nuclear war they fought, and also referred to Huguenin's theory that flying saucers were Atlantean aircraft which were brought to the Subterranean World prior to the occurrence of the catastrophe that sank Atlantis. The abandonment of their former home on top of the four-sided sacred mountain in the center of Atlantis (Mount Olympus or Meru, later memorialized by the four-sided, truncated pyramids of Egypt and Mexico) and their skyward journey over the Rainbow Bridge of the Aurora Borealis, through the polar opening, to the new home in Walhalla, the golden palaces of the city of Sham-ballah, capital of Agharta, the Subterranean World. This migration of the Atlantean god-rulers to the Subterranean World, prior to the destruction of Atlantis, was referred to in Teutonic mythology as the "Gotter-dammerung" or Twilight of the Gods. They made the journey in flying saucers, which were Atlantean aircraft.

Whereas, in the days of Atlantis, flying saucers flew in the Earth's outer atmosphere, after they entered the Subterranean World they continued to fly in its internal atmosphere in its hollow interior. After the Hiroshima atomic explosion in 1945 they rose again to the surface in numbers, seeking to avert a nuclear catastrophe. The tragedy that befell Atlantis was due to its scientific development running ahead of its moral development, resulting in a nuclear war, which heated the atmosphere, melted polar ice caps and brought on a terrific deluge that submerged the continent. A group of survivors, led by Noah, found refuge in the highlands of Brazil (then an Atlantean colony), where they constructed subterranean cities, connected by tunnels to the surface, to prevent destruction by radioactive fallout and flood.

According to Plato's account, Atlantis was submerged by a series of inundations which came to a climax about 11,500 years ago. Some four million inhabitants lost their lives. Those who were more spiritual and were fore-warned escaped in time to Brazil, where, it is claimed, they or their descendants still live in subterranean cities.

In this connection it is interesting to refer to Jules Verne's book, "A JOURNEY TO THE CENTER OF THE EARTH," which presents a similar conception of the earth's formation as did Gardner's book by a similar name. Verne describes a party of explorers who entered a volcanic shaft, and after traveling for months, finally came to the hollow center of the earth, a new world with its own sun to illuminate it, oceans, land and even cities of Atlantean origin. Verne believed that prior to the destruction of Atlantis, some of the Atlanteans escaped and established subterranean cities in the earth's hollow center. Since most of Verne's predictions were later verified, it is possible that this one also will be— not by entering a volcanic shaft, but by an aerial expedition through the polar openings into the hollow interior of the earth.

One of the early German settlers in Santa Catarina, Brazil, wrote and published a book in old German, dealing with the Subterranean World, deriving his information from the Indians. The book described the Earth as being hollow, with a sun in its center. The interior of the earth was said to be inhabited by a disease-free, long-lived race of fruitarians. This Subterranean World, the book claimed, was connected by tunnels with the surface, and these tunnels, it was claimed, open mostly in Santa Catarina and surrounding parts of South Brazil.

The author has devoted nearly six years to investigations to study the mysterious tunnels which honeycomb Santa Catarina, obviously built by an ancient race to reach subterranean cities. Research is still in progress. On a mountain near Joinville the choral singing of Atlantean men and women has been repeatedly heard— also the "canta gallo" (cock crowing), which is the standard indication of the existence of a tunnel opening leading to a subterranean city. The crowing is not produced by a living animal but probably by some machine.

The Russian explorer, Ferdinand Ossendowski, author of "Beasts, Men and Gods," claims that the tunnels which encircle the earth and which pass under the Pacific and Atlantic Oceans, were built by men of a pre-glacial Hyperborean civilization which flourished in the polar region at a time when its climate was still tropical, a race of supermen possessing scientific powers of a superior order, and marvelous inventions, including tunnel-boring machines we know nothing about, by means of which they honeycombed the earth with tunnels. We shall now quote from Ossendowski's remarkable book relating his own experiences in Mongolia, where belief in the existence of a Subterranean World of Agharta, ruled by the King of the World, who resides in his holy city of Shamballah, is universal. Ossendowski writes:

" 'Stop!' said my Mongol guide, when we crossed the plateau of Tzagan Luk, 'Stop!'

"His camel bowed down without the need of him ordering it. The Mongol raised his hands in a gesture of adoration and repeated the sacred phrase:

"OM MANI PAEME HUM."

"The other Mongols immediately stopped their camels and began to pray.

" 'What happened?' I wondered, bringing my camel to a halt.

"The Mongols prayed for some moments, then mounted their camels and rode on.

" 'Look,' said the Mongol to me, 'how the camels move their ears with terror, how the manes of the horses remain immobile and alert and how the camels and cattle bow down to the ground. Note how the birds stop flying or the dogs barking. The air vibrates sweetly and one hears a song that penetrates to the hearts of all men, animals and birds. All living beings, seized with fear, prostrate themselves. For the King of the World, in his subterranean palace, is prophesying the future of the peoples of all the earth.'

"Thus spoke the old Mongol.

"In Mongolia, with its terrible mountains and limitless plateaus was born a mystery which was preserved by the red and yellow lamas. The rulers of Lhasa and Ourga guarded this science and possessed these mysteries. It was during my trip to Central Asia that I heard for the first time this Mystery of Mysteries, to which I formerly paid no attention, but only did later, when I was able to analyze it and compare certain testimonies frequently subjected to controversy. The old men on the border of Amyil told me an old legend, according to which a Mongolian tribe, seeking to escape from Genghis Khan, hid in a subterranean land. Later, near Nogan Lake, I was shown by Soyota a door which served as the entrance to the kingdom of Agharta. It was through this door that a hunter entered into this region and, after he returned told of his visit. The lamas cut off his tongue to prevent him from speaking about the Mystery of Mysteries. In his old age, he returned to the entrance of the cavern and disappeared into the Subterranean World, which memory always brought emotion to the nomad.

"I obtained more detailed information from Hou-touktou Jelyl Djamsrap de Narabanch Kure. He told me

the history of the arrival of the all-powerful King of the
World to the door of exit of the Subterranean World,
his appearance, his miracles and prophecies. I then commenced to understand this legend, this hypothesis, this
collective vision, which, no matter how we interpret it,
conceals not only a mystery but a real force which
governs and influences the course of the political life of
Asia. From that moment, I commenced my investigations. The lama Gelong, favorite of Prince Choultoun
Beyli, gave me a description of the Subterranean World.

"More than six thousand years ago, he said, a holy
man disappeared into the earth accompanied by a tribe
of people and never returned to its surface. This inner
world was also visited by various other men, as Cakya-
Muni, Undur-Ghengen Paspa, Baber and others. No
one knows where they found the entrance. Some say it
was in Afghanistan, others say it was in India.

"All inhabitants of this region are protected against
evil, and no crime exists within its boundaries. Science
developed tranquilly, uninterrupted by war and free from
the spirit of destruction. Consequently the subterranean
people were able to achieve a much higher degree of
wisdom. They compose a vast empire with millions of
inhabitants governed by the King of the World. He
masters all the forces of nature, can read what is within
the souls of all, and in the great book of destiny. In-
visibly he rules over eight hundred million human
beings, all willing to execute his orders.

"All the subterranean passages in the entire world
lead to the World of Agharta. The lamas say that all the
subterranean cavities in America are inhabited by this

people. The inhabitants of submerged prehistoric continents (Lemuria and Atlantis) found refuge and continued to live in the Subterranean World.

"The lama Turgut, who made the trip from Ourga to Pekin with me, gave me further details: The capital of Agharta (Shamballah) is surrounded by villas where live the Holy Sages. It reminds one of Lhasa, where the temple of the Dalai Lama rises on top of a mountain surrounded by temples and monasteries. His palace is surrounded by the palaces of the Gurus, who control the visible and invisible forces of the earth, from its interior to the sky, and are lords of life and death. If our crazy humanity will continue its wars, they may come to the surface and transform it into a desert. They can dry the oceans, transform continents into seas and cause the disappearance of mountains. In strange vehicles, unknown above, they travel at unbelievable speed through tunnels inside the earth. The lamas found vestiges of these men in all parts and in inscriptions on rocks; and saw remains of the wheels of their vehicles.

"When I asked him to tell me how many persons visited Agharta, the lama answered: 'A great number, but most of those who were there maintain the secret as long as they live. When the Olets destroyed Lhasa, one of their regiments, in the mountains of the southwest, reached the limits of Agharta and were then instructed in mysterious sciences, for which reason the Olets and Talmuts became prophets. Certain black tribes of the east also entered Agharta and continued to live there for centuries. Later they were expulsed from the Subterranean World and returned to live on the surface of the earth, bringing with them knowledge of the mystery of prophecy by means of cards and reading the lines of the hand. (They were the ancestors of the gypsies.) In a certain region in the north of Asia there exists a tribe which is on the verge of disappearing and which frequents the caverns of Agharta. Its members can invoke the spirits of dead which live in space.'

"The lama then remained silent some time and then, responding to my thoughts, continued: 'In Agharta, the sages write on stone tablets all the sciences of our planet and of other worlds. The Chinese Buddhist sages know that well. Their science is the most advanced and purest. In each century the sages of China united in a secret place near the sea and on the backs of a hundred large turtles that come out of the ocean they write the conclusions of the divine science of their century.'

"This brings to my mind a story that was related to me by an old Chinese attendant in the Temple of Heaven in Pekin. He told me that turtles live for three thousand years without air or food and for this reason all the columns of the blue Temple of Heaven rest on the backs of living turtles, so that wooden supports would not rot.

"Many times did the rulers of Ourga and Lhasa send ambassadors to the King of the World, said the lama librarian, but they could not reach him. However, a Tibetan chief, after a battle with the Olets, came to a cavern whose opening bore the following inscription:

" 'THIS DOOR LEADS TO AGHARTA.'

"From the cavern left a man of beautiful appearance, who presented to him a Golden tablet bearing strange inscriptions, saying:

" 'The King of the World will appear to all men when comes the time of the war of the good against the evil; but this time has not yet come. The worst members of the human race have yet to be born.'

"Chang Chum Ungern sent young Prince Pounzig as an ambassador to the King of the World. The ambassador returned with a letter for the Dalai Lama of Lhasa. He wished to send him a second time but the young ambassador never returned."

Chapter X

CONCLUSION

From the evidence contained in this book, confirmed by many Arctic explorers whom we cite, we come to the following conclusions:

1. There is really no North or South Pole. Where they are supposed to exist there are really wide openings to the hollow interior of the Earth.

2. Flying saucers come from the hollow interior of the Earth through these polar openings.

3. The hollow interior of the earth, warmed by its central sun (the source of Aurora Borealis) has an ideal subtropical climate of about 76 degrees in temperature, neither too hot nor too cold.

4. Arctic explorers found the temperature to rise as they traveled far north; they found more open seas; they found animals traveling north in winter, seeking food and warmth, when they should have gone south; they found the compass needle to assume a vertical position instead of a horizontal one and to become extremely eccentric; they saw tropical birds and more animal life the further north they went; they saw butterflies, mosquitoes and other insects in the extreme north, when they were not found until one is as far south as Alaska and Canada; they found the snow discolored by colored pollen and black dust, which became worse the further

north they went. The only explanation is that this dust came from active volcanoes in the polar opening.

5. There is a large population inhabiting the inner concave surface of the Earth's crust, composing a civilization far in advance of our own in its scientific achievements, which probably descended from the sunken continents of Lemuria and Atlantis. Flying saucers are only one of their many achievements. It would be to our advantage to contact these Elder Brothers of the human race, learn from them and receive their advice and aid.

6. The existence of a polar opening and land beyond the Poles is probably known to the U.S. Navy in whose employ Admiral Byrd made his two historic flights and which is probably a top international secret.

Chapter XI

UFO'S OR FLYING SAUCERS IN
ANCIENT TIMES...

Did Super Beings From Space Ever Visit Earth?
Classical Writers Reported So.

Each Age interprets unusual events in the language of its own experience, whether it be Ezekiel describing sky objects in the symbology of angels and precious jewels, or Monk Lawrence in A.D. 776 marveling at flaming shields from heaven spitting fire at the Saxons besieging Sigiburg, or modern men speculating the Unidentified Flying Objects are of extra-terrestrial origin.

Now that astronomers blazon the belief that life exists throughout the universe, speculation naturally exists that spacemen could have landed on Earth in ages past.

Is there evidence?

For more than 2,000 years it was recorded by nearly all the greatest intellects of Greece and Rome although most of the records of antiquity have been destroyed. In the surviving Classics there is ample evidence of UFO's and probable extra-terrestrial interventions.

Our theologians dismiss the ancient Gods as anthropomorphisms of natural forces, as if entire races for hundreds of years would base their daily lives on lightning and thunderbolts! Yet logic suggests that the old Gods of Egypt, Greece, Rome, Scandinavia and Mexico

were not disembodied Spirits or anthropomorphic sym-
bolisms but actual spacemen from the skies. It seems
that after the great catastrophes remembered in legends,
the "Gods" withdrew and henceforth have been content
merely to survey the Earth, except for an occasional
intervention in human affairs.

Apollodorus wrote, "Sky was the first who ruled over
the whole world," surely signifying domination by space
beings. The Roman Emperor Julian vowed, "We must
believe that on this world . . . certain Gods alighted."

Aeschylus, Euripides, Aristophanes, Plautus and Me-
nander frequently introduced a "Deus ex Machina" (a
God from a Machine) to untangle the plots of their
plays.

Aristotle, Plato, Pliny, Lucretius and most other phi-
losophers believed that the Gods were supermen living
in the realms above.

A century ago a German grocer Heinrich Schliemann,
using the Iliad as a guide, defied the ridicule of the pro-
fessors and dug up Troy. Can we dig up records of
spaceships in other classics?

Following are some examples from the works of
ancient writers, scrutinized for UFO references:

B.C. 498: Visitations

". . . Castor and Pollux were seen fighting in our army
on horseback. . . Nor do we forget that when the
Locrians defeated the people of Crotona in a battle on
the banks of the river Sagra, it was known the same day
at the Olympian Games. The voices of the Fauns have
been heard and deities have appeared in forms so visible
that they have compelled everyone who is not senseless

or hardened to impiety to confess the presence of the Gods."

—Cicero, Of the nature of the Gods, Book I, Ch. 2

B.C. 325: Visitations

"There in the stillness of the night both consuls are said to have been visited by the same apparition, a man of greater than human stature, and more majestic, who declared that the commander of one side and the army of the other must be offered up to the Manes and to Mother Earth."

—Livy, History, Book VIII, Ch. 11

B.C. 223: Bright Light, Three Moons

"At Ariminium a bright light like the day blazed out at night; in many portions of Italy three moons became visible in the night time."

—Dio Cassius, Roman History, Book I

B.C. 222: Three Moons

"Also three moons have appeared at once, for instance, in the consulship of Gnaeus Domitius and Gaius Fannius."

—Pliny, Natural History, Book II, Ch. 32

B.C. 218: The Sky Is Filled

"In Amiterno district in many places were seen the appearance of men in white garments from far away. The orb of the sun grew smaller. At Praeneste glowing lamps from heaven. At Arpi a shield in the sky. The moon contended with the sun and during the night two moons were seen. Phantom ships appeared in the sky."

—Livy, History, Books XXI–XXII

B.C. 217: Fissure in the Sky

"At Falerii the sky had seemed to be rent as it were

with a great fissure and through the opening a bright light had shone."

—Livy, History, Book XXII, Ch. 1

B.C. 214: Men and Altar

"At Hadria an altar was seen in the sky and about it the forms of men in white clothes."

—Julius Obsequens, Prodigiorum Libellus, Ch. 66

B.C. 163: An Extra Sun

"In the consulship of Tiberius Gracchus and Manius Juventus at Capua the sun was seen by night. At Formice two suns were seen by day. The sky was afire. In Cephallenia a trumpet seemed to sound from the sky. There was a rain of earth. A windstorm demolished houses and laid crops flat in the field. By night an apparent sun shone at Pisaurum."

—Obsequens, Prodigiorum

B.C. 122: Three Suns, Three Moons

"In Gaul three suns and three moons were seen."

—Obsequens, Prodigiorum, Ch. 114

B.C. 91: Gold Fireball

"Near Spoletium a gold-colored fireball rolled down to the ground, increased in size, seemed to move off the ground toward the east and was big enough to blot out the sun."

—Obsequens, Prodigiorum, Ch. 114

B.C. 85: Burning Shield, Sparks

"In the consulship of Lucius Valerius and Caius Marius a burning shield scattering sparks ran across the sky."

—Pliny, Natural History, Book II, Ch. 34

B.C. 66: From Spark to Torch

"In the consulship of Gnaeus Octavius and Gaius Suetonius a spark was seen to fall from a star and increase in size as it approached the earth. After becoming as large as the moon it diffused a sort of cloudy daylight and then returning to the sky changed into a torch. This is the only record of its occurrence. It was seen by the proconsul Silenus and his suite."

—Pliny, Natural History, Book II, Ch. 35

B.C. 48: Thunderbolts, Visitations

"Thunderbolts had fallen upon Pompey's camp. A fire had appeared in the air over Caesar's camp and had fallen upon Pompey's . . . In Syria two young men announced the result of the battle (in Thessaly) and vanished."

—Dio Cassius, Roman History, Book IV

B.C. 42: Night Light, Three Suns

"In Rome light shone so brightly at nightfall that people got up to begin work as though day had dawned. At Murtino three suns were seen about the third hour of the day, which presently drew together in a single orb."

—Obsequens, Prodigiorum, Ch. 130

B.C.?: Suns, Moons, Globes

"How often has our Senate enjoined the decemvirs to consult the books of the Sibyl! For instance, when two suns had been seen or when three moons had appeared and when flames of fire were noticed in the sky; or on that other occasion when the sun was beheld in the night, when noises were heard in the sky, and the heaven itself seemed to burst open, and strange globes were re-marked in it."

—Cicero, On Divination, Book I, Ch. 43

A.D. 70: Chariots in the Sky

"On the 21st of May a demonic phantom of incredible size . . . For before sunset there appeared in the air over the whole country chariots and armed troops coursing through the clouds and surrounding the cities."

—Josephus, Jewish War, Book CXI

A.D. 193: Three New Stars

". . . three stars . . . suddenly came into view surrounding the sun, when Emperor Julianus in our presence was offering the Sacrifice of Entrance in front of the Senate House. These stars were so very distinct that the soldiers kept continually looking at them and pointing them out to another . . ."

—Dio Cassius, Roman History, Book LXXIV

A.D. 217: Visitation

"In Rome, moreover, a 'Spirit' having the appearance of a man led an ass up to the Capitol and afterwards to the palace seeking its master as he claimed and stating that Antoninus was dead and Jupiter was now Emperor. Upon being arrested for this and sent by Matermainus to Antoninus he said, 'I go as you bid but I shall face not this emperor but another.' And when he reached Capua he vanished."

—Dio Cassius, Roman History

The above references are only a sampling of the evidence available. Consider just five writers: Julius Obsequens recorded 63 celestial phenomena; Livy, 30; Pliny, 26; Dio Cassius, 14; Cicero, 9.

Romans fervently believed that two strange horse-
men, taller than normal men, alike in age, height and
beauty, saved the day for Posthumus at Lake Regillus
and, that same day, miraculously appeared in the Forum,
announced the victory, and departed forever.

A contemporary historian described two shiny shields
spitting fire around the rims, diving repeatedly at the
columns of Alexander the Great in India, stampeding
horses and elephants, and then returning to the sky.

When we recall that Romulus was borne to heaven by
a whirlwind while giving judgment on the Palatine Hill,
that his successor Numa Pomilius, used magic weapons,
that Livy, Pliny the Elder, and Julius Obsequens tell of
mysterious voices, celestial trumpets, men in white garments hovering in airships, several suns and moons together, sudden new stars, and superhuman apparitions
descending among men and then vanishing, we suddenly
feel we are reading the wonders of the Bible.

By some strange twist of the human mind, we worship
prodigies in old Palestine as manifestations of the Lord,
yet scoff at identical phenomena occurring at the same
time only a few hundred miles away.

Evidence exists; all we need to do is examine it.

Chapter XII

FLYING SAUCERS, PROPULSION AND RELATIVITY

Solve the UFO propulsion problem and you open the whole universe to man! Here's a theory that may explain it.

For the past few years we have been visited by large numbers of foreign space craft. Actually these visits probably have been occurring for a long time; perhaps for what we call geological time periods. However, in 1947 or shortly before, the number of visits rose sharply. Since 1947 a great number of persons around the world have seen the famous flying saucers, or unidentified flying objects (UFO's).

Over the past few years observers have watched the craft perform acrobatic maneuvers of an astonishing nature. Apparently most of the saucers do not depend on any propulsion familiar to our science or, at least, familiar to us until recently. Only a very few have been reported with propellers, and while some have reaction motors, either jets or pure rockets, many do not even have these. Indeed, the typical flying saucer floats above the earth with no visible means of support and then dashes off at a truly breathtaking speed to some other part of the globe.

The lack of any known propulsion system capable of such effects has led many persons to speculate that the owners of the saucers have been able to master the physics of gravitation. The propulsion system used must in some way apply what is popularly called anti-gravity. There is hardly any way, at least so far as both laymen and experts can see, how their ability to stay above the earth with neither jets, propellers nor extensive lifting devices can be explained. But a further, though closely related, enigma is the typical saucer motion. For not only has gravity been conquered, but inertia seems to have been conquered also. Many reports—some of them apparently authentic—tell of UFOs suddenly appearing in the sky from nowhere and then disappearing, seemingly in an instant. Unless some optical trick is involved, the saucers must be capable of truly extraordinary acceleration. Typical of saucer reports, as they appear in the local presses throughout the world is the object seen cruising along at a few hundred miles per hour and then, suddenly, seen to dash away at what must be thousands of miles per hour.

In addition to these extraordinary linear accelerations the saucers seem to outwit inertia in other respects. At very high speeds they appear to make perfect right angle turns and even reversals of direction, without disastrous results to their structure or their crew—if these exist. At least two of my friends have told me of seeing flying saucers, moving through the sky at very high speeds, make instantaneous right angle turns.

Still another good trick—they seem able to move through the atmosphere at rates of speed and at levels of air density which clearly are incompatible with any pub-

licly known technology. As an object moves through the
air the friction of the molecules striking its surface causes
the material to heat. In our very fast jet interceptors cool-
ing systems are necessary. We all know how meteors entering the earth's atmosphere, and nose cones of missiles
re-entering the earth's atmosphere, heat to such a point
that in many cases they disintegrate or burn up completely. Yet moving at comparable speeds in a denser atmosphere, UFO's do not seem to show these effects. To
be sure, luminosity often appears about them—especially
at night—and occasionally trails of smoke vapor appear,
but the machine itself seems to survive. To missilemen
this is most curious.

At stake, in all these maneuvers, is our understanding
of the stubborn laws of inertia which govern our world.
Newton first formulated these clearly in his double
principle that an object at rest tends to remain at rest
unless a force is applied, and if a force is applied it tends
to take motion in the direction of the applied force and
proportionally to it. These Newtonian laws of inertia still
are the basis of much of our scientific world view. But
combining them with the known molecular binding
forces of matter, which are equally fixed in the order of
nature—at least so we think—makes the saucer's behavior very difficult to explain.

When the flying saucers accelerate from 0 speed to
many thousands of miles an hour in a few seconds, why
isn't their internal machinery torn apart and any crew
member squashed?

Anyone who has taken a curve at too high a speed

knows the persistent tendency of his vehicle to continue along the original line of motion against the force of his tires and steering mechanism.

Similarly when a flying saucer makes a sudden turn, traveling many thousand miles an hour, why don't the molecules or crystals of its metallic structure literally tear apart—from the great strain imposed by the laws of inertia?

And finally, as the saucers rush through the atmosphere why don't the molecules of the atmosphere, striking against the saucer cause heat through friction and eventually burn the object up?

It is these very remarkable performances that have led many persons to believe the saucers are not real. Material objects cannot behave this way! The saucers must be moving light, optical illusion, mirage, diffraction pattern, atmospheric lens or, to PFO's (Persons Farthest Out), ghosts or spirits.

The head of Air Force Intelligence remarked rather wistfully after the great Washington Airport sightings some years ago that he (i.e. the Air Force) did not have anything with infinite energy and no mass. Any person trained in non-relativistic physics believes it would be impossible for ponderable mass to behave as the UFO's behave.

However, the trouble with this argument seems very real, indeed. For saucers do exist! They have been photographed! They return firm radar images! And at close range they look very much like craft made of metal or transparent materials similar to plexiglass. Aside from their unusual tricks they seem to have all the characteristics of hard material objects which are designed, fabricated, manufactured, or what you will.

If the saucers are real solid vehicles we must revise our ideas of nature in one of two respects. Either we must conclude that our knowledge of the rules which hold atoms and molecules together is incomplete, or we must revolutionize our concept of inertia. If both alternatives were beyond the reach of modern science there would be no reason to prefer one over the other.

But, in fact, there is a perfectly good way of explaining the saucers within modern physical theory. To do so, however, we must pass to the abstract heights of physics, in particular to Albert Einstein's General Theory of Relativity. Now, before you are too frightened, let it be said that the General Theory is not as complex and intricate as some persons think. Its reputation for difficulty arises from the fact that, to grasp it, a transvaluation in the way we feel about the world is necessary.

Newton's concept of inertia tells us that an object stays in its place unless some force is applied to it and when the force is applied the object moves with the force. Newton had rather mixed ideas of why inertia exists. At one point in his *Principia* it is almost inherent in matter. At another point inertial or centrifugal forces arise from something called absolute space. The persistence of matter in its state, according to Newton, comes from its relation to an absolute world of space more final than any material system we can think of.

This notion of Newton's was never satisfactory and in the last part of the 19th Century the Austrian physicist and philosopher Ernst Mach turned his critical mind to it. Mach, whom we all know for his Mach numbers of aerodynamics, was also a forerunner of the Vienna Circle which developed logical positivism. To him anything beyond observation—such as absolute space—was unreal. Hence he proposed that inertia was a reference to *all the matter in the universe*. By all the matter in the universe he meant all the fixed stars, or in our day, when we realize that the cosmos is made up of vast numbers of stars collected in vast numbers of galaxies, to all the galaxies. For Mach an object subject to the laws of inertia was relative to all the stars, or as we would say today, all the nebulae.

Yet Mach's principle, as Einstein called it, had a difficulty. It did not supply any physical link between the stars and an inertial system. Mach just substituted the universe for Newton's absolute space as a system of coordinates in which objects existed and moved. He did not take us any further down the road to showing what inertia is, or why it works the way it does.

Perhaps we should say, rather, that he took us a little way and he took Albert Einstein a very long way.

In 1916 Einstein proposed his General Theory of Relativity. In effect it was a theory of universal gravitation *and inertia*. Einstein reduced the two forces *to the same thing* and expressed this in his famous Principle of Equivalence: gravitational and inertial forces are indistinguishable and equal. His illustration of this is a man in an elevator deep in space. The man is away from any large objects. If the elevator is moving uniformly at any constant speed, from a very small one to a very large one, the man will seem quite weightless. He will sense no motion or gravity pull. However, if the elevator speeds up, if it is pulled by its cosmic cable along the direction of the man's height at an increasing speed, the man will begin to feel as if gravity is acting upon him. When a certain acceleration is reached, equivalent in earth's measurements to 32 feet per second, the man will imagine that he is back on the earth and is being pulled down by the earth's gravity just the way he was before he left earth. Actually, of course, he is not. His false impression is merely the result of inertia and the acceleration of his elevator. There is no gravitation or, more correctly we should say, there is no large object in his vicinity.

Thus Einstein illustrated the fact that inertia and gravity have exactly the same effects on the observer and cannot be distinguished on the basis of local observations.

He went further. He sought to explain gravity and inertia in the same physical terms. While the weight of objects on a large celestial body like the earth is caused by the latter's gravitational attraction, the inertial behavior of objects is explained by the gravitational attraction of all matter everywhere. To use a simple analogy, the pipe resting on the table in front of me remains where it is largely because all the stars and nebulae of the cosmos are pulling on it, and they are pulling on it in all conceivable directions. It is as if a million million million little wires were attached to the pipe symmetrically all around it and are pulling it equally at the same time in every direction. Similarly, as I throw my pencil across the room it goes in a straight line (aside from earth's gravity) because it is being pulled at every right angle to the direction of its flight by the totality of matter in the universe, by all the stars or nebulae. Thus inertia in the familiar world is really gravitation but not the gravitation of the earth or of any single big body near us, but the gravitation of every particle in the universe; it is the sum effect of gigantic push, pull, or field depending on how you regard the still elusive gravitational mechanism.

But how, you ask, does this help us explain how flying saucers fly?

If the owners of the saucers have been able to devise a revolutionary means of anti-gravity, say an electromagnetic screen which they put around their craft, this will mean that as the earth's gravity is overcome the gravity-inertia of all the rest of the universe will be overcome also. If the gravitations or ultra particles or fields which account for the gravitation of the earth are screened out the gravitational effect of the rest of the universe will be screened out also. Thus the saucers, with their anti-gravity screen, will be able to fly above the earth and they will be able to ignore the laws of inertia. They will be literally floating in a little cup or envelope where neither gravity nor inertia play any role. If the creatures who have built and man the saucers have mastered gravity they must, according to Einstein, have overcome inertia, also.

The key to the rather strange thing I have just said is to think how an atom or a molecule, or a group of them which make up an object, will behave if no inertial influence can reach them. The pipe on my desk, now at the slightest touch of my finger, may fly across the room. Similarly, if I now throw my pencil across the room the slightest breeze will send it off at a right angle toward the other side of the room. In other words, we may assume that the atoms and matter in an inertia-free area will become almost totally free in their environment. They can move in one direction as easily as in another. They have no tendency to remain in the rigid envised position which inertia would ordinarily hold; they can fly away freely in any direction in which a slight force impels them.

I think this explains how the saucers can accelerate from zero to thousands of miles an hour and decelerate at the same rate, how they can engage in the dramatic maneuvers reported. Once a force, of whatever kind, impels them in a direction different from their line of movement, there is no tendency for their atoms and molecules to continue moving in their former direction. Thus, there is no strain upon the structure of the ship and the molecular binding forces of its material are not torn apart. Again, its occupants, if they can live in such an inertia-less world, are not crushed in the slightest or even disturbed by the gyrations of the superstructure around them. Presumably they could sit quietly reading a book without knowing that their craft actually was doing the most remarkable acrobatics.

The concept of a gravity-inertia screen may also explain why the saucers do not burn up as they speed through the atmosphere. Consider a molecule or atom of gas as bumping along against other atoms in the atmosphere, subject to the laws of inertia as everything else is, but not causing very much damage or disturbance because it has little mass; a saucer rushes by and the molecule finds itself within the gravity-inertia screen. Suddenly this little air molecule is entirely free! It no longer carries kinetic punch; it can bump into anything without causing the slightest friction. In other words, it enters the screen like a bullet and strikes the saucer like a feather.

However, as the saucer rushes on, this molecule of air pops out the back of the screen in a very agitated state. It is now again in the inertial world and starts bumping into other highly agitated molecules. Its tiny little punch is magnified as a result of the friction which was not possible and this causes a release of energy—the luminosity seen about the saucers, especially at night.

At this point perhaps we should review what we have said and what we have not said.

In a sense, we have explained how the saucers fly but we have not explained how the gravity-inertia screen is generated. Sometimes flying saucers under observation during the day through polaroid glasses, and some photographs of saucers, exhibit a kind of halo or corona about them. Of course, this well may be a physical token of the screen. However, the way it is produced is still a mystery, at least to this writer.

It is almost certain that in some way the field involves electricity and magnetism—for the effects of both have been noticed in connection with saucers. It is also likely that nuclear energy is used in the generation process, because increases in radioactivity background levels also accompany UFO flights. But of the exact mechanisms which produce the screen we know nothing. Research in this area is highly classified. The earth power which first develops the technique will have an immense military advantage. It may render not only aircraft, but ballistic missiles obsolete.

Let us consider what man's mastery of gravity and inertia may mean for his life on earth and his progress in space—if other races allow him to make any. In the first place, down here on earth the control of both gravity and inertia may well transform much of our economic system. We can think immediately of gravity-free airplanes plus the advantages of being able to control the inertia which governs (and hampers) so much of our lives.

If inertia can be controlled a five-year-old child can bounce an elephant upon its knee; the work of the world may be done with tiny amounts of energy—depending, of course on how much is needed to produce the gravity-inertia screen. We may be able to move mountains with only the quantity of electricity to light a house. The whole phenomenon of friction may be within our range of manipulation; railroad trains may be able to rush down their tracks covered with an inertial screen driven by only fractional horse-power motors.

The idea of inertia-free flight opens up interesting possibilities for space travel. Given inertia-free flight, space may no longer be a barrier to solar travel!

Some astronomers and physicists, pointing to the enormous amounts of energy required to accelerate even a tiny payload near enough to the speed of light to make the journey to the nearest star in any reasonable period of time, have held the view that the only communication mankind will ever have with intelligent life elsewhere in the galaxy is by radio.

The distances between stars are measured in light years and only a limited number of stars are within one-half the light year equivalent of four score and 10. Thus the necessity for approaching the optical velocity in interstellar travel becomes obvious. Yet, even to approach it under the old law of inertia is a difficult matter; some scientists believe it is impossible.

Dr. Frank Drake illustrates the problem by calculating that to deliver the *Encyclopedia Britannica* to our nearest stellar neighbor would require such a huge rocket that its blast-off would incinerate the entire state of Florida.

Other scientists, of course, have believed that interstellar travel is possible, even under the limitations of an inertial world. The great German physicist, Professor Singer, once proposed an inter-stellar vehicle capable of sweeping up the hydrogen atoms in space in a gigantic net and converting them into fuel along the way.

But if we are able to develop a gravity-inertia screen we may be able to approach the optical velocity with very little energy actually required.

It also may mean that higher species, who long ago discovered the same technique, have voyaged back and forth between the stars quite regularly. This would, in turn, increase the likelihood that our solar system is visited by races from other stars.